So You Want to Be in

Show Business

Zorro television series, 1958.

So You Want to Be in
Show Business

A Hollywood Agent Shares the Secrets of
Getting Ahead Without Getting Ripped Off

STEVE STEVENS SR.
with **John D. Cady**

CUMBERLAND HOUSE
NASHVILLE, TENNESSEE

With all this horseshit, there must be a pony around here somewhere!

—James Kirkwood

Published by
Cumberland House Publishing
431 Harding Industrial Drive
Nashville, Tennessee 37211

Cover design: Stacie Bauerle
Interior design: Mary Sanford

Library of Congress Cataloging-in-Publication Data
Stevens, Steve, 1939–
   So you want to be in show business : a Hollywood agent shares the
secrets of getting ahead without getting ripped off / Steve Stevens,
with John D. Cady.
       p. cm.
   ISBN 1-58182-453-X (hardcover : alk. paper)
   1. Motion pictures—Vocational guidance. 2. Television—Vocational
guidance. I. Cady, John D., 1966– II. Title.
   PN1995.9.P75S74 2005
   791.4302'8'023—dc22

2004029833

Printed in the United States of America
1 2 3 4 5 6 7—11 10 09 08 07 06 05

I dedicate this book to my soulmate, Rosemary, who stuck with me through the ups and downs over the past thirty years. To my two sons, Steve Jr. and Mark, who allowed me to share their worlds and learn what it was like to grow up all over again. To my dad, who supported me in everything I tried to do. When he passed on, I realized that I will always be someone's child. To my mom, who never got a chance to see life in the right perspective, but in her own little world loved us very much. To those clients who were so loyal throughout the years, and the many who became good friends. To my Special Olympics buddies, who help me realize that, if you try, you can do almost anything. And last but not least, to JC for answering so many prayers.

—STEVE STEVENS SR.

Contents

Acknowledgments

I wish to thank the following people:

Wende Doohan, Ken Guran, and my dear wife Rosemary for helping me gather and scan the photos used in this book.

Tommy Cole, Emmy-winning makeup artist and former Mouseketeer, not only for contributing to this book, but for being a dear friend for over forty years.

Cynthia Smalley, the talented photographer, for taking time out of her busy schedule to contribute a chapter for this book.

Gary Marsh, owner of Breakdown Services, who wrote a chapter for this book. He helped change the way agents now function, and has made a great contribution to our business.

John Cady, who kept me honest and not self-serving. Early on he made me realize that my knowledge should be passed on to those who might need it. Find out about John at www.lyrichyperion.com.

Ken Beck, author of *The Cowboy Cookbook*, who introduced me to the wonderful people at Cumberland House.

Mary Sanford for her terrific editing skills and wonderful insight to what I have lived and written.

Dave Jamison and Seth Tucker for their proofreading and editing contributions and encouragement.

To those bigger than life show business greats who each in their own way taught me the *DOs* and *DON'Ts* of my chosen profession: Chuck Connors, George C. Scott, Bobby Darin, Stanley Kramer, Al Ruddy, James Doohan and all those others who let me tag along.

INTRODUCTION
Who Is Steven R. Stevens Sr.?

A father; a husband; a man dedicated to his family, team roping, and, yes . . . show business; but those are just the broad strokes. Right now, I'm someone who wants to share my years of experience with a new generation of artists and professionals who want to become a part of the business that has shaped my life.

Coming from a show business family myself, I broke into films (almost literally) at the age of thirteen. With over two hundred acting credits, including starring in the "Annette" series on *The Mickey Mouse Club*, Walt Disney's *Zorro*, *Playhouse 90*, *Mr. Novak* (regular), *Gunsmoke*, *The Roy Rogers Show*, and as Betty Hutton's son on the *That's My Mom* TV series, I feel like I know the acting game. But there's so much more to a life in this business than merely appearing in front of the camera. If you let it, and if you are professional and play from the heart, show business will reveal itself as one of the most rewarding fields you can pursue. This is the secret I want to share with you.

At the age of twenty-three, after a stint with the U.S. Marine Corps, I decided to give up acting and become an agent—not a secret agent, but the kind that helps other actors realize their dreams. In a few short years I was representing such stars as Jack Palance, Bobby Darin, Chuck Connors, and Slim Pickens, just to name a few. But my wanderlust didn't end there.

In 1969 I decided to explore another aspect of show business. I became a casting director at a company called Screen Gems. Right away, I was saddled with the responsibility of casting two major TV series, *Nancy* and *The Partridge Family*, as well as the movie-of-the-week *The Feminist and The Fuzz* with Donna Mills and David Hartman. After a year, however, I began to feel stifled and found that casting did not satisfy my creative desires in the way that I needed it to. Screen Gems had purchased a series idea that I wrote for Johnny Cash called *The Hard Luck Circuit*, and though it was never produced, it paid the bills for a long time. I requested a release from Screen Gems, which gave me the opportunity to take off to New Mexico for some soul-searching in the high desert.

Much of this time I spent writing country songs and studying the customs, wisdom, tradition, and art of the Native American Indian. My search was cut short by a call from a friend—apparently George C. Scott wanted to meet me and I was urged to return quickly to Hollywood. Mr. Scott was looking for someone to cast his first directorial effort (a feature called *Rage*), in which he would also star. This turned out to be a propitious encounter: I would eventually go on to cast all of the films George C. Scott produced or directed.

But at the time, I have to admit, I was scared to death. I mean, George C. Scott, the legendary star of *Patton*, wanted to meet me? I shouldn't have been nervous, though. George was the kindest, most sincere man I have ever met. We really hit it off, and I went on to collaborate with him not only on *Rage*, but also on *Oklahoma Crude* and *The Savage Is Loose*. Other credits I garnered during my return to casting included *The Longest Yard, Enter the Dragon, Murph the Surf,* and the three-hour television pilot for *How the West Was Won*.

In 1976 I returned to what has proven to be my most enduring passion—I opened my own agency. Many of my clients have been with me for decades. The Stevens Group has contributed to the careers of the following talents, among many others:

Frankie Avalon

Noah Beery Jr.

Amanda Blake

Bruce Boxleitner

Foster Brooks

Chuck Connors

Bobby Darin

Clarence Gilyard

Alan Hale ("The Skipper")

Richard "Jaws" Kiel

Johnny Knoxville

George Lazenby (James Bond)

George ("Goober") Lindsey

Doug McClure

Jack Palance

Slim Pickens

Denver Pyle

Guy Williams

Hal Williams

Michael T. Williamson

Also *Star Trek* stars Nichelle "Uhura" Nichols, James "Scotty" Doohan, Walter "Chekov" Koenig, and George "Sulu" Takei.

Through it all, I have maintained a special relationship with the individuals I've encountered and the business as a whole. It's a relationship that has not only empowered me as an individual, it has shaped my entire life. It is this relationship, the relationship I have with show business as a *business* and so much more, that I invite you to share.

So You Want to Be in

Show Business

Breaking In

My Way

Today, security at movie studios and independent productions is as tight as it is at the White House. The days of sneaking onto the lot to drop in on casting directors or have lunch in the commissary and hobnob with the rich and famous are long gone. Just ask Christopher Richard Hahn:

> Spielberg's publicist frowned upon Hahn's attempt to make contact. "Obviously if anybody gets onto a studio lot illegally, he's going be stopped," said Marvin Levy. "We don't want to give anybody else ideas of doing the same thing."
>
> —*E!, February 7, 2002 (Josh Grossberg)*

It's not at all like the days when I was a teenage actor. After selling out my newspapers on the most infamous corner in the world (Hollywood & Vine), I would walk to Eagle Lyon Studios, where I would climb through a small hole in the massive wall. Eagle Lyon Studios is now a shopping mall on Santa Monica Boulevard in West Hollywood. It's right next to The Lot, which was Samuel Goldwyn Studios back then (one of the top four at the time). It later became Warner Hollywood, then Spelling Studios, then Samuel Goldwyn again but only as a rental facility, and now it's known as "The Lot."

Even back in the 1950s, studio guards carried guns. So I knew I would have to be very careful. I walked through the lot with a Leo Gorcey swagger (he was from *The Bowery Boys;* nowadays you'd probably call it a "Christopher Walken swagger") and looked like I belonged. I was all of thirteen years old. It was a Saturday, and in those days they worked a six-day week. On the set of the very first anthology show, Frank Wisbar's *Fireside Theatre,* I was spotted by a for-real, fin de siècle German director, complete with boots and a riding crop. He pulled me aside and gave me a speaking role. That was it. At thirteen I was a professional actor and a member of the Screen Actors Guild.

I sincerely doubt that could happen today. There is no Schwab's Drugstore where legend suggests Lana Turner and several other actors were "discovered." The days when being an extra was a proud profession and there was a chance you'd be pulled out by a top director and given a nice role in a film because he "loves your face" are as long gone as Andy Griffith and the mythical Mayberry.

So I decided to write this book for the beginning actor. It's a look at the business from my point of view, sprinkled with *DOs* and *DON'Ts,* and filtered through a thorough understanding of the business called "showbiz." I have honed this view for over fifty years as an actor, producer, casting director, and agent. It is unique, but it is universal. These are the things you really need to know to "make it" in Hollywood, and I don't mean just find yourself in front of a camera, I mean find your true self, your happiness, and your place in the world.

As a teen actor my credits include: *The Mickey Mouse Club* (with Annette in her first series), *The Roy Rogers Show, That's My Mom* (as Betty Hutton's son), *Gunsmoke, Playhouse 90, Climax,* and *High School Caesar,* just to name a few.

As a casting director, my credits include four years with George C. Scott, casting everything he directed on film, as well as *How the West Was Won* (three-hour pilot), *Murph the Surf, The Longest Yard* (not credited), the first season of *The Partridge Family,* and more.

I have executive produced two feature films, *Bugbuster* and *Knockout.* And for over thirty years I have been a SAG franchised agent, and very proudly so. I have represented struggling actors who never made it, struggling actors who did, top movie and television stars on their way up, and some on their way down. The knowledge I am sharing with you is intended to help you avoid mistakes I see people in the business making every day, particularly neophyte actors. I'd like nothing more than to help you make a living at your chosen profession, to help you find that profession, and to see you find success and happiness in show business. Enjoy!

It All Begins with a Dream

"Mommy, I want to be on the TV!" an eight-year-old child blurts out with all the innocence and sincerity in the world. The lucky parents are able to distract and subtly dissuade the child from this dreaded course. The unlucky ones get ready for the parade: singing lessons, dancing lessons, acting lessons, grammar school plays, middle school plays, high school plays, finding a college with a good drama program, leads in college plays, and, and, and . . . finally . . . the ultimate decision is made. The now "grown-up" child will move to Hollywood and pursue that all-too-common, lifelong dream—to be on TV.

Uncle George comes up with a bus ticket. Mom and Dad have put away a few thousand to help with the move and "getting started." The old VW doesn't have enough left in its used-up body to make the trip, so that bus ticket comes in real handy. After three days and no sleep in antic-ipation of becoming a movie star, you find yourself on one of the most notorious streets in the world.

Vine Street just south of Hollywood Boulevard in Hollywood boasts a dirty bus depot, right next to a not-much-cleaner McDonald's. "Hey there, can you spare some change?" is the customary welcome to the City of Angels. You notice there are several panhandlers working the pas-sengers getting off the bus. You wonder whether any of these poor, sad, dirty, stoned souls ever got off a similar bus full of hopes and dreams, in search of a career in show business. I don't know, but I *can* tell you that they will make more money in a year at their current profession than 70 percent of the members of the Screen Actors Guild make from acting. Not to discourage you, but this is the kind of thing it's handy to know before you get on the bus. So if you're already here, or still coming anyway, let's get down to the business end of show business and figure out how to make that dream of yours work.

DO!

The first thing you need to do is find a place to live and get a cell phone if you don't already have one, so people know where and how to reach you. There are rental guides for free in the news racks on almost every corner. Get one. It's probably the last thing you will get for free until you make the Hollywood "A-list."

It used to be that you would be advised to avoid areas with too much graffiti, even if the rent was cheap. However, nowadays graffiti might be a sign that property values are about to explode—it's art, you know. At any rate, the "work," as they say, of an actor is looking for work, and you will be spending more time than you can imagine transporting yourself from audition to audition, so something central is crucial. With traffic being what it is, it might even be the key to your sanity. Hollywood, once on the "avoid" list, is popular and central. Other good central areas include Culver City, Studio City, Los Feliz, Silver Lake, and Echo Park. If you do choose the relatively graffiti-free and affordable areas in the San Fernando Valley, understand that you will be spending hours on the freeway for each audition.

Once you've settled in, you need to get a *Thomas Guide* (at any bookstore or gas station) so you can find your way around town. There are very few studios in the Hollywood area and you will soon realize that your dream will be that much harder to achieve without a car. Maybe you should have taken a chance on the old VW? Oh well, *shoulda, coulda, woulda.*

DO!

Buy *Back Stage West.* It comes out every Wednesday and is the best trade paper for the beginning actor. It has a great quantity of information about independent productions, student films, and workshops. Yes, workshops are a great place to make new friends, enemies, contacts, and learn how to use your acting skills Hollywood-style. A good workshop can also take the place of group therapy. Hey, don't laugh, I'm talking from experience. Plus, it's an easy way to find a friend in what can be an exceedingly lonely place.

DO!

You have the right to audit an acting school or workshop. This is very important. Do not make your decision based on how many hot babes or macho studs look available or appear as lonely as you feel. Does the coach or teacher have a personality that gels with yours? Do they seem approachable? Do you feel comfortable in the surroundings? All of these, in addition to the relative availability of babes or studs, should inform your decision. Try to be happy with at least three of the four criteria. Once you are settled in a class or workshop, start working on your photos. Ask some of the other students if you can look at their pictures, and ask your coach/teacher if he or she can recommend a photographer.

DO!

Back Stage West has hundreds of ads for photographers looking for your business. They are priced from under two hundred to thousands of dollars. Set up an appointment and review their work/book. Photographers, like actors, have their own style. One might take great photos, but might not be the right feel for your persona. Some will be better for pretty people and others for characters. Some will only shoot inside and others will prefer working exclusively outside. Some will charge for a makeup person, and others will throw it in for free. It can get very expensive, as it's not just taking your photos, it's getting hundreds reproduced. Your theatrical agent is going to want different pictures than your commercial agent. "Oh God, this is stressing me out!" Don't let it. I'm going to walk you through the process.

DON'T!

Do not rush into this. Your photos are a very important tool for you and your agent and managers. They must be absolutely right. So take your time investigating what is out there and at what price. You'll be surprised at how many options are at your disposal.

DO!

Call UCLA, USC, Cal State Northridge, Loyola Marymount, Pasadena Arts Center, or any other college you can find in the phone book. Ask for the photography department; you're almost sure to find students there that will do your shots for the experience. You might even make a lifelong friend. Offer yourself as a model in exchange for free pictures. Anything you can do to save money and network will help you toward your goal.

DON'T!

Do not pose nude in exchange for photos. This can come back to haunt you when you're starring in a hit television series or when you consider running for office. I will go into the subject of nudity in more depth later (see Chapter 15).

DO!

Your agent and manager are going to be selling you, and they will want to be part of the process of packaging and imaging (and therefore picking the images with which they will be plying their craft). If your agent, new agent, manager, or acting teacher does not like your pictures—no matter how good you think they are—respect that. They know what they are doing.

Take a generic picture or a good character shot. Print up as few copies as possible and use them to send to prospective business partners. You might even find someone in your workshop who has a camera and will take a few shots for you if you pay for the film. Remember that the people working with you will want to manipulate your image, so avoid committing to anything before they've had their say.

Now that you have a few photos, you need to make up a résumé. Remember, everyone has to start somewhere. You would be surprised how much you actually do have to put on your résumé that just might impress someone. Your education, plays, any modeling, all those sports, and don't forget that you know how to drive a tractor or a stick-shift. But believe me, don't lie about any of your credits or special skills. If you say you can ride a horse or rock climb, you had better be able to—not only could you get hurt, you could embarrass your agent. You do not want to embarrass your agent!

Send your picture and résumé to all the student films and non-union productions listed in *Back Stage West*. Audition for anything and everything you can. This is a great way for you to learn the interviewing and auditioning process before you meet the big guys. Make your mistakes where it will do the least damage, and learn, learn, learn, learn, learn (while you're at it). Remember that some of today's student directors asking you to volunteer your time will be tomorrow's A-list directors and producers deciding multimillion-dollar budgets.

Trade Papers and Advertising

Can They Help?

I am a great believer in advertising. I had an uncle who owned a furniture store. He took out a full-page ad every Sunday, letting the public know who he was and what he had to sell. I was only eight or nine, but I thought that was very interesting. I asked him why he took out all those ads in the paper. "If I have a couch for a good price, and someone is looking for a couch with a good price, they know where to go to find that couch with a good price." I never forgot that and have used that theory throughout my acting and agent careers.

Now, in the marketplace of show business there are many ways to go. Let's get into the more expensive and popular way to advertise, the "trade papers." I think it is tacky to take out any ad in *Variety* or the *Hollywood Reporter* that isn't at least a full page. Of course, it is very expensive, and if I could only do one it would be in *Variety*. They are both considered the godfathers of show business trade papers, but *Variety* is larger and I feel you get more for the buck.

Throughout the years I have found it easier to get blurbs in *Variety* without a publicist than in the *Hollywood Reporter*. *Variety* also seems more actor-friendly. When my wonderful dad, a former song-and-dance man with many acting credits, passed away, I sent his obituary to both trades, and only *Variety* printed it. For me that was the final straw.

When you take out a small ad in either trade paper stating that you are appearing or guest-starring on a television show, it looks as though you didn't make or don't have enough money for the full-page ad. Many have disagreed with me about this over the years, but I stand by my belief. Also, if there is any way the ad can be in Friday's edition, that's the best day. It has the production schedule and people are more apt to take the Friday trades home.

A more reasonable way of advertising when you have something airing, or a play production, or even have new representation, is oversized postcards. They are effective, but do not deliver near the impact and range of the trades.

The reason you want to purchase the trades, which should always include *Back Stage West*, is so that you know what is going on out there. If you are in L.A., New York, or some little town in Georgia, and you are interested in how you can make a living in show business, you need these papers. *Back Stage West* is essential for the beginning actor, but the other two list all the movies that are going to be shot around the United States complete with when, where, and by which production company. *Back Stage West* lists the non-union productions, college productions, music videos, and theater productions of all sizes. You are allowed to submit yourself for these projects. It also has wonderful articles about casting directors and agents, and very good insight into the people that you might be meeting somewhere down the road.

I read all the trades I can, every day. It keeps me abreast of what's going on in the business of which I am lucky to be a part. Heck, what I can't find in the trades I have mentioned, I can sneak off and read in the latest issue of the *Enquirer*. Happy reading!

Amanda Paytas. (Photos by Cynthia Smalley)

Candice Rose. (Photos by Cynthia Smalley)

Photos and Résumés

Finding Your Persona

Your pictures are extremely important. Often they will be the only thing you've got working for you in the initial stages. Whether they get you called in or not makes all the difference. Later, in the section on keeping your agent happy, I will detail my particular views about headshots. Before we look at what I've done with some of my clients, let's hear from a good friend of mine, Cynthia Smalley. Cynthia is an excellent photographer and really captures that extra something that can put you over the edge (or get you through the door). She specializes in portraits, commercial photography, pets, family, pregnancy, and more! (You can see her work at www.smalleyphoto.com.) She'll be able to tell you more about what the photographer wants and needs than I will.

Cynthia was featured in *Cosmopolitan* magazine as one

of their "Top Ten Fearless Females of 1998." As *Cosmopolitan* reported:

> Eight years ago, Cynthia Smalley was another struggling actress. But when she became frustrated over not being able to get a natural-looking headshot, Smalley switched focus. With a few pointers from other photographers, Smalley taught herself photography, and today, at thirty-four, she single-handedly juggles successful studios in San Francisco and Los Angeles and a thriving volunteer life. She is an integral part of a charity devoted to introducing inner-city kids to the fresh air of ranch life. Smalley and her crew teach the kids to ride horses, clean stalls, and enjoy the open range.

Cynthia says that she loves "meeting those tough kids, with all their attitude. Then seeing their faces light up when they get a horse to kick up its hooves."

Her photography style displays a great deal of creativity, knowledge, and experience, and her images are true works of art. The following is Cynthia's advice, and then I'll check in afterward and we'll look at some real headshots.

So you're sitting with your photographer, planning your new headshots. Let's talk about some marketing ideas.

Think about what "type" you are and what characters you would play. How would you realistically be cast in a TV commercial, TV show, or film? Come up with ideas of your own and bring them to the table. Ask your agent how he or she will be submitting you for auditions. How your agent sees you being cast is important. Hopefully you are both in agreement on this; it should play an important role in how you see yourself working in the industry.

Some agents like to have the specific "types" you can play individually shot, so you have several different 8 x 10s to apply to the needs of any given audition. For example, a woman might have shots for a newscaster/lawyer, mom, nurse, cop, glamorous/upscale, etc.; and for men: CIA/FBI, cop, lawyer, dad, coach, etc.

To get your own ideas for this, look through magazines at the pictures of actors. From Vanity Fair to Men's Journal to Cosmo, all of these magazines have ideas in photo form that you can tear out and bring to your agent and photographer. These will help you describe what you like visually. You really want to make sure everyone is on the same page, so to speak, with marketing your career. They can also give you ideas about clothing and hair and body positioning. If you find a certain way of sitting or standing is comfortable and really feels like "you," then do it in front of a mirror to see how it reads. Practice observing without self-judgment or criticism the best angles of your body and face, so you will know how to sit and move a little more easily when the shoot is in progress. Of course your photographer should be helping with this as well, but it's nice to feel some sense of it yourself.

If your agent just wants one or two pictures of you, for example a more serious theatrical shot and a lighter, more playful commercial shot, then clothing will be a simple project. Always choose wardrobe that has a simplicity to it so that you will shine through. Distracting patterns take one's eye away from the person in the picture. Choose necklines that are attractive on you. Pick a material and style that you are comfortable in and truly like. It will naturally convey a part of you. The camera reads, and likes, textures and tactile feel in the fabric of your clothing, for example: suede, corduroy, linen, or denim.

I prefer shooting either lighter or darker colors for a look with more contrast in the printed shot rather than all medium and gray tones, which don't pop out as nicely. Always wear colors that look good on you and your skin tone; even though it is black-and-white film, you can still tell the difference in the shots.

Now, when you're choosing clothing for character types, find out what these types of people wear in the most detail you can. If you could play a trucker, but you aren't one and you're not sure what one would wear, then tune into a TV show or rent a movie that has a trucking theme. You might also go to a coffee shop at a truck stop and do your own research. You'll be able to spot the "uniform," and you may even spot the perfect "ED'S TRUCK STOP AND BAIT" baseball cap for sale. Pick one up for your shoot and head to the local thrift shop for the rest of the "look." If you need a military, nurse, or diner waitress outfit, there are costume rental businesses in most major cities that have great authentic wardrobes you can rent and sometimes even have tailored to fit you. Ask your agent or photographer, or consult the Yellow Pages and the Web.

Do come prepared top to bottom and make sure that everything is pressed and clean (if it should be) when you arrive at your photographer's location.

Your hair is important! It is important to know your hair, and how to wear and style it. If you are a wash-and-wear kind of person, then find a hairdresser that can give you the cut and the help in simple styling that gets you that look consistently.

If you're more complex in your hair needs, the more you can do to learn the "How to's" the better off you will be. Even when booked on a film or commercial, the hair people hired may be on their first job or not at all familiar with your particular hair type. Things like this can and will happen. But you can still look great on the job, even if you need to fix yourself up, if you've pursued this information. The same is true for makeup, both for men and women. Know your face! Know how to do the basics at least. This career encompasses a wide variety of facades of the self; the more you know, the further ahead you are at the start. It may be wise to bring your own small makeup and hair kit to every shoot, just in case.

Arrive with a clean face. This means with no flaking skin that needs to be exfoliated. Your eyebrows should be shaped and your upper lip waxed (if needed) on women. For men who are shooting with beard growth, make sure the cheek and neck lines of definition are clean (if that's the look) and bring an electric razor with batteries for the clean-shaven look if required. Men may want to keep a styptic pencil in their makeup kit for stopping any blood flow from nicks.

Let the makeup person know your basic makeup prefer-ences. If freckles are your trademark, let them know you'd like them to show. If you go to auditions with lots of makeup, almost no makeup, or none at all, let them know this so that they can customize your general look and still do what's needed for the camera and lighting. When your makeup is finished, take a very good look and trust your intuition. Be true and politely speak up if you feel a part, or the whole of it, doesn't look right on you. It's a learning process discovering the right looks for your type and what makeup looks like in person and on camera. I find that makeup for black-and-white film is not that different from makeup for color film; however, what you see on your face will seem somewhat less distinct in black and white.

Now that you have chosen your photographer, makeup artist, and hair person; decided on a marketing plan for your niche in the industry; and your wardrobe, hair, and makeup are all perfect, what can you do to present these uniquely individual and inspiring parts of yourself to the greatest advantage during the shoot? How can you put into still pictures all the skills that you've honed for putting into moving pictures? The acting techniques you utilize on the stage or in film are just as essential to your photo session. What do you do to evoke joy, sensuality, or your edgy side? If you can't do this easily, get into an acting class right away. Don't underestimate the acting that will be required in your photography session, and be prepared to work.

Whatever your acting style, there are some basic human elements that need to be evoked during your shoot. Physical sensation is important. Feel your hands, feet, and breath. Get in touch with the sitting or standing, leaning or lounging, or whatever your body is conveying, and make sure it is in alignment with the look you are trying to achieve.

A strong emotional choice is paramount. Whatever you are feeling in your gut and body and heart when a photo is taken is what the photo will evoke to a person looking at it afterward. So if you are pushing for something that's not there, and you are uncomfortable in your body and feeling fake, ask the photographer to take a minute. Collect your thoughts and feelings, and think about something that is heartfelt, makes you passionate, or makes you feel mischievous. Use that thought to connect you to the parts of your personality that are engaging.

For the individual being photographed, it is often internal, mental work that creates an appropriate state of mind, mood, or sensuality. Some people, such as dancers, respond to movement as a way of freeing themselves to the moment. Other people make acting choices that release the emotional body to a particular momentary state of being. Whatever it takes to release, it is this moment of being that you are hoping to capture on film. Practice techniques that work for you to free you into these moments.

Here are some examples of how I've worked with individuals and "the moment":

1. You're out with someone very special having dinner. As you both are talking, laughing, exchanging animated ideas, or just lovingly listening, you notice in your mind's eye what your body is doing. Check in with your posture. Are you leaning comfortably forward, shoulders down, breathing nicely with an interested, connected, and perhaps even playful look in your eye? What does this feel like? Close your eyes and remember; see if you can create just two moments of this or some other personal experience. Get very specific while in front of the camera.

2. I was photographing an actor who had played football for the 49ers. He was having a hard time bringing true joy to the shoot, until he picked up his football. He squeezed the ball as if he was going to throw it, and he became alive with realistic enthusiasm. At this point, I got a great shot—minus the football.

3. A beautiful opera singer who looked like a swan but appeared timid, unexpressive, and cool was posing for promo shots. I talked with the young woman, trying to find a way into a passionate part of her persona. When I

learned that the woman (a former ballet dancer) still loved to dance, I had her point her toe in first position. That simple movement revealed the singer as a beautiful, sensual woman whose past expertise could now flow through her. She transported herself into a physically different way of being, and I captured that on film.

4. An actor's agent wanted headshots of the actor looking like an intelligent, calculating doctor/lawyer type. The actor, who was terrible at math, purposely tried doing the twelve times table in his head, knowing that he would have to concentrate intensely to do it. The look on his face was of a strongly focused, intelligent man and resulted in the perfect shot.

5. The octogenarian and consummate talent James Doohan, Scotty from the original Star Trek, was doing promo shots as a spokesperson for a computer company. In addition to being very professional, he is humorous and personable. He revealed to me that he was having a baby with his wife of twenty-five years, whom he loved madly. When we discussed this, he beamed an adoring smile (radiating real warmth) at the thought of his wife and baby. From the moment we shared this connection, the shoot went beautifully and exuded real emotion.

Listen, whatever brings a sense of life to you and your body, use it in the shoot. No one needs to know what it is . . . even a secret can be great!

Here are some basic tips to think about. Do your work and let the photographer do his or her work. Try not to question and "produce" the photographer's side of the work. You will have much more fun, and better photos, if you try to stay inspired and connected while simply revealing those darling quirks that make you unique and castable. Think of the shoot as just one in a lifetime of photo sessions. In each one, you will learn something new, get a different perspective on what you can bring out of yourself, and discover a new angle on yourself—one that you like and endorse. And most of all, try not to criticize yourself and your efforts. Just give yourself a pat on the back for being brave and showing up. I know of an acting coach who says this is 80 percent of getting a job. The other 20 percent, in my opinion, is being prepared and sending thank-you cards to everyone who helps you.

• • •

Now take a look at these pictures of a few of my clients. You'll get an idea of what to do.

Cliff Emmich needed new headshots and his photographer was out of the country. I took these photos with Cliff's camera.

What I needed from senior actress Peg Stewart.

Showcases

Keeping Your Cool in the Chute

*C*asting director and agent showcases are put on by companies that charge the actor and pay the casting directors and agents.

In today's marketplace, showcases are a necessary evil. Like everything else, there are good ones and those that are a total waste of time.

Casting directors do not have the time to meet actors on general interviews, so how the hell are they going to get to know you and your talent if they don't see you perform? Their time to prep each week for a television show is very limited. Production meetings with their producers and directors, pulling submissions, and pre-reading actors doesn't leave much time for generals.

This is where showcases come into play. Beware, they can become very costly. They can run $30 to $45 for each one, depending upon how far up the food chain the casting director is. Some larger-scale "industry" showcases that promise a smorgasbord of agents, managers, casting directors, *et al.* can be as much as $700 to $800. I think it is important for the beginning actor to do as many of these showcases or workshops as possible, but at the same time to pick and choose wisely. This is not only for the beginning actor. Consider the case of a good comedic actor who has built a reputation as a hard-edged dramatic actor. The showcase presents an opportunity to show off the undiscovered side of your talents.

Not all casting directors are casting the type of project that will fit you. In those cases, it is a waste of your money and time. Do some research on the casting directors you are planning to see. Check around. Have they ever brought someone in from a showcase? Some of my clients have been very lucky and have picked up several jobs from showcases and workshops. One client booked all six episodes of a new series from a workshop. Another client guest-starred on a top television show because the casting director loved her work in a showcase and brought her in to read for the director. Of course, these were actors ready to work who had more than paid their dues.

Some showcases do not offer you the chance to choose your partner, and at some workshops the casting directors bring their own material. You could be partnered up with a real shlub or draw some poor material and have a bad experience. It's just luck of the draw. You could also draw someone more advanced than you and have that person really show you up despite your having turned in a good performance. Go over the list of casting directors offering workshops with your agent or manager. They should be able to point you in the right direction.

I am sometimes a little skeptical of assistant casting directors. Most have not paid any dues and yet run entire workshops. They use the name of their employer and the show or movie that the employer is casting to get actors to come. But do they have an eye for talent and can they get an unknown actor in for an audition? There are a few questions that need answering. However there are some up-and-coming assistant casting directors that I deal with all the time. They are very talented and it will only be a question of time before they have their own company.

My main problem with CD workshops is the attitude I have witnessed in some of the casting directors, and the truly awful attitudes exhibited by many assistant casting directors. The process frequently degenerates into a power trip instead of an opportunity to help someone get through a trying time in pursuit of their dreams. Besides, they are getting paid for this by the actors who are auditioning for the jobs they might be able to distribute, and that alone creates some degree of compromise.

Be very careful with agent workshops. These can be self-serving ego trips for the agents. Like CD workshops, there are good ones and bad ones. Don't waste your time with agents from the bigger agencies. You'll get lost in the shuffle. Read up on the agencies in *The Agency Book,* and do your homework. Know what the agent is about and how you fit into his or her needs. There are some agents who are looking for new talent and do find it at these workshops.

DON'T!

Please do not do a workshop or showcase if you are not ready. I have seen such bad performances with unprepared talent trying scenes beyond their abilities. I do not understand why some of these workshops do not audition actors more carefully. A little selectivity in the talent presented would go a long way. Bad acting makes me nauseous! Again, here is a potential risk: the companies that will let any actor in at a price might not show you in a good light. Even if you stand out a little, you will be judged by the company you keep. Be judicious in your showcase and workshop choices.

Meeting with Casting Directors

And the All-Powerful Assistant Casting Director

Hmm, casting directors . . . I've been one! And now I work with them every day. They have a difficult job to do, but the really good ones have such an eye for talent it will amaze you. With any luck you'll be meeting a great many of them, so it's important to keep track of which ones you know and which ones have seen your work. Remember what you wore, what the part was for, what you did, and how you did it. Taking notes on this stuff after every meeting is a very good idea. The more you can find out about this mysterious and essential craft and those who practice it, the better.

Always remember that the casting director really wants you to do well. They are looking for a person to come in there and knock their socks off, so they can take your picture to the director or producers and say, "This person is perfect, absolutely what we wanted." In turn, the producers and directors want you to repeat your success. They would love it if you filled them with confidence that they had done their job and needn't spend any more time on this important preliminary work. So don't be shy, give them what they want. When you walk into an audition situation, remember that everyone in the room is hoping you'll do well.

DO!

Help your agent by doing theater and casting-director workshops. Then let your agent know what workshop you did or what casting directors came to your play. When choosing a casting director's workshop, narrow it down and do some research. You will not fit into the type of television show or feature every casting director is casting. If you do not fit the bill for *NYPD Blue* or *The Shield*, don't waste your time with those casting directors. If comedy is not your bag, don't waste your time with a casting director working on sitcoms.

DON'T!

Do not do any casting director's workshop until you are ready. You can really blow it! Many of the better casting directors have memories like an elephant. That is why they are on top of the food chain. Five years later they will remember that you were awful that rainy night in November five years ago.

Assistant Casting Directors

My personal opinion is that too many assistants put on the workshops. Some are up-and-comers, but most couldn't get their brother a meeting with their own boss for the opening in the coffee-retrieval department. It's mostly big attitude and little talent. However, some assistants play a big part in casting shows and features, and some even do it all without receiving the credit. I have BTDT (been there, done that). So before you shell out your hard-earned money, investigate the CDs and assistant CDs. Work with your agent on this; usually he or she will know who would be good for you. Also, there are only four or five companies that get the top casting directors to come, so you don't have far to look.

How NOT to Meet a Casting Director

I do not recommend the following methods! In Atlanta, while filming *The Longest Yard*, I had my casting office in my hotel room. Maids at the hotel would constantly put pictures of their friends and relatives under my pillows and sheets. Although I was on the sixth floor, small rocks were thrown at my window to get my attention. Pictures were routinely slipped under my door, and a waiter even put his headshot on top of my steak and bought me several rounds of drinks. A cabbie took me to a workshop run by his mother, instead of where I had wanted to go. Could this be called kidnapping? All of these tactics are not effective.

In Denver, while casting the movie/pilot for *How the West Was Won*, I took a bathroom break after several hours of meeting cowboys and Indians. As I sat there contemplating my next few hours of casting, the door to my stall flew open (I guess I forgot to lock it). In front of me stood a lanky, forty-something cowboy in boots, spurs, and hat, the complete cowboy ensemble. "Mr. Stevens, my name is Luke. I can ride, I can rope, I can shoot a gun. Here is my portfolio." He handed me his oversized portfolio as I sat there in shock. "Excuse me, but I'm a little tied up right now, go to the casting office," I replied to this cowboy from another place and time. I then noticed the expression on his face change. "Oh, God. I'm sorry. I was so excited to have a chance at being in a Western movie . . . I didn't mean to disturb you on the shitter." He grabbed his portfolio back, and I never saw him again.

Some people are overwhelmed by the fantasies they create regarding breaking into the business. They see the business as a change or transformation, and exaggerate what it's going to mean for them to become a _____. It's hard to explain, but it's very important to let go of the idea of "being discovered" or "making it overnight." Just work hard, train, do theater. If you're an actor, act wherever and whenever you can. If you're working in any other aspect of the business, work on your craft and perfect your skills. Don't assault people in their private lives, let your work show, and keep at it.

Getting an Agent

And Keeping an Agent

You are now ready to pursue an agent. Don't worry, it's not as bad as it seems. If you play your cards right you may be on the verge of making a lifetime friend. On the other hand you may "only" meet someone who can teach you a great deal about the industry. You *can* sign with the wrong agent, however, so be careful. Here are some guidelines to help you through this precarious and absolutely essential process.

DO!

Buy updated labels of all the agents. You can get this from Breakdown Services (www.breakdownservices.com) or the Samuel French book store on Ventura Boulevard in Studio City (11963 Ventura Boulevard, Studio City, 91604; 818-762-0535) or Sunset Boulevard in Hollywood (7623

Sunset Boulevard, Hollywood, 90046; 323-876-0570). If all goes well you will be spending a great deal of time here, so figure out which one is nearest to you. They are great stores in which to browse, and have an unbelievably extensive catalogue of show business–related information. *The Agencies* is a great book to get started with; it will help you focus on who might be taking on new clients or representing your type. If you're at the Studio City location, take a walk down Radford Street as you leave. It is now CBS Radford, a television studio, but years ago it was Republic Pictures. Roy Rogers and Gene Autry chased the bad guys up and down this western street (which unfortunately is no longer there). *Gunsmoke* was shot on this lot for over twenty years. If you close your eyes, you can still hear the hoofbeats and the voices of all those who went before you. This lot is full of the ghosts and great memories of Old Hollywood. On the other hand you might just bump into the cast of *Will & Grace*, who currently film in the studio.

DO!

Send a picture, résumé, and cover letter to all the agencies with which you feel a connection. Look particularly for those agencies that work with or are looking for your type. Try to be as objective as possible in your self-assessment. It would be nice to just show up and *be* Julia Roberts, but you have to start somewhere. A good place to start is with an agent who needs your type and with whom you can work and learn.

DON'T!

Avoid sending cover letters with any of the following phrases:

1. *"I am happy with my agent but I'm looking for someone who can take me to the next level."*
   This indicates you're using your current agent and will use your next, and your next, as rungs in the ladder. An agent will see through a letter such as this and toss your picture. Whatever happened to loyalty? Build a relationship with your agent and honor it.

2. *"I am looking for an agent that will fight for me, and have the same passion and love for [my career, the art, or the craft] that I have. I need someone who will fight for me at every turn."* (These are just a few of the quotes I receive all the time.)
   This turns an agent off, particularly if you aren't a member of SAG. If an agent decides to work with you, it's because they feel they can make money with you. It might be for the short term or the long term; it might be in movies, television, or commercials; but I'll bet you a year's subscription to *Back Stage West* it's not for the passion or love you might possess. It's strictly about dollars and cents.

DON'T!

Don't send postcards for the first contact. They are a waste of time and money. You can't cram a photo, résumé, and eye-catching note on a 4 x 6-inch card. If you have already met with an agent and you are now SAG or have something exciting to relate, a card with a photo is okay.

DON'T!

Do not call an agent and ask if they received your submission three weeks ago. I constantly get calls from actors who sent their submissions weeks, even months, ago and are just "following up." Please, you know agents receive hundreds of submissions a week. If you didn't, you know it now. If there is an interest, an agent will rush to get you in for a meeting. There is no need to be pushy or have an attitude—you get more with sugar than vinegar (or even strong espresso).

DON'T!

Don't drop by or drop off your submission if there is a sign on the agent's door telling you not to do it. Climbing through a window or approaching an agent in a restroom is also a big no-no and might even get you arrested.

DO!

Try to get an agent to come to a showcase you put on, or do some of the more legit agent showcases around town. Be careful though, as there are many of these that are not worth your time and money. My son and partner, Steve Jr., has picked up several new, untried, non-union actors from commercial workshops he has attended. He is extremely happy with the outcome. I have signed a few new actors from some of the theatrical workshops I've done. Two have worked out, and the others . . . well that's a whole different chapter.

DO!

When you are given the opportunity to have a meeting with an agent, be prepared! Have all your pictures in a portfolio or stacked neatly when you hand them to the agent. Dress up, not down. I have met so many sloppy and unwashed actors in my day. Maybe James Dean could get away with it, that was his persona. If you have his unique combination of talent and luck, it might work for you too. But I find it an immediate turn-off. While we are on the smell factor, if you smoke, wear clothes from the cleaners and shower. Do not have a cigarette for at least two hours before your interview. I am a former smoker and couldn't care less, but if you smell up my office I will conclude that you will smell up a casting director's office and potentially limit your salability. Do not offend casting directors! People have varied and sometimes radical opinions about smoking and it's not worth taking a chance. For some people the stench of smoke translates into an immediate goodbye.

DON'T!

When you are in a first meeting with an agent, and he asks you to tell him a little about yourself, don't start with, "I was born in rural Podunk and moved to suburban Who Cares City. I've always wanted to be an actor." What the agent is asking is what have you been doing to progress your career. Don't talk about the extra work you've done or how much you might "want it"—those are negative. Don't talk about what genres interest you, or how you see yourself, or whose career you admire or desire. "I'm essentially focused on features at this point in my career; I'm a solid, vulnerable performer who would do well in the sort of atypical roles Johnny Depp usually lands," will get you a quick introduction to the door.

Don't knock the agent you are currently with, if you

are lucky enough to have one. That's a big negative. It's a very small world, and you don't know who knows who. If you have been blessed or cursed with representation, don't treat your agent as the enemy. Don't call and bug your agent, and never put your agent on the defensive. That is most definitely a negative. It should be a partnership, and like any relationship it takes time to progress. You are not the only one on your agent's roster. If not you, your agent has a talent pool of thousands of actors to choose from. They're all out there, and most of them would kill or die for an agent—any agent.

DO!

Get the kind of pictures your agent wants. Don't argue with your agent regarding this matter. Your agent is the one selling you, and needs the kind of ammunition he feels comfortable using. The type of photos being used for commercial and theatrical are not much different in today's market. I hear all sorts of things like, "You only need one good photo," or "Why should I get a picture in a cowboy hat?" Blah, blah, blah . . .

Listen, you can come up with any excuse not to have several pictures, and they are all wrong. If you are what I call an "Ivory Soap"–looking actor (pretty/handsome) and you fall into that one area, then okay. If you are a character actor and can do many things, then damn it, bring the pictures to your agent so he can help you meet your goals!

DON'T!

Now, don't take a picture with a cowboy hat standing next to a horse unless you can ride that horse. Don't take a picture wearing a leather jacket and sitting on a Harley unless you can ride that hog. Go ahead, laugh. I have been doing this for over thirty years and the only thing in this business that hasn't changed is the type of pictures an actor needs to weed out a little competition, so if you do know how to do those things, take the pictures to emphasize it!

Here are a few examples: three clients, all top guest-star actors, all around sixty and looking upscale. Two have pictures in a general's uniform. They get out every time I submit them for an officer role. The client without that type of picture (who, by the way, has better credits than the other two) doesn't get called in for those roles as much. I have five older character actors. Two have a farmer picture, and they always get out for that kind of part. The other three do not—and one of them happens to be a real farmer. It's not who you are, it's who they think you are.

Remember, I was a casting director myself, and I know that sometimes casting does not have time to be overly imaginative. Why not help them along? Two clients have pictures in cop uniforms and they get out for that role every time we submit them. Are you getting the idea? Don't forget that it's no use getting these types of pictures unless you can back it up with your talent. As I was writing this, sure enough, a client got an interview for a feature playing a small-town sheriff. Can you guess which photo I submitted? Yes, the one in the small-town sheriff's outfit, badge and all. I will not waste my time on a client that doesn't get me the pictures I request. Life is too short.

DO!

Take care of your agent at Christmastime. This is a business. You don't have to spend a great deal of money. Bake some cookies or something. It's a way of saying, "Thank you for what I think you have been trying to do and I hope you continue to do." Just because it hasn't happened, it doesn't mean your agent isn't working hard for you. Every time you're submitted for a role it takes time and money. If you don't work, your agent doesn't get paid for the work done. Also, take care of the secretary and the assistant—someday they could be the boss, or even running a network.

DO!

Do agency workshops. There are several who are really looking for new clients and not showing up just for the money and coffee. Yes, they get paid to be there. I have done some, and Steve Jr. does them. Beware the self-serving, trying-to-seem-important, been-an-agent-for-a-year, dispensing-wrong-information-to-young-actors-who-need-to-learn type of agent. (They drive me crazy!) Take it all with a grain of salt, and throw out the crap that doesn't apply to you. Let the young agent with the big ego do the workshop with the assistant casting director with no background and a serious attitude. I might pay to sit in on that for laughs, but you should do your research and avoid it.

Just because these people see themselves as being in a position to help you doesn't mean you should endure their crap. Your time will come. After a workshop of any kind, follow up with a nice picture and résumé. Stay on this every few months as you progress.

DON'T!

Don't waste your time doing agents' workshops for the bigger agencies either. You will more than likely be lost, overlooked, and forgotten in an agency concentrating on multimillion dollar deals for Brad Pitt and Julia Roberts. You'll probably be wasting your money, and if you did happen to get signed you might even be hurting your career. No one there will have the time or the inclination to work on your $1,800 industrial or regional California Fruit Growers Association commercial.

Leaving Your Agent

When

It takes time to get things moving when pursuing a career. You are not the only client your agent has, although you may wish you were at least on top of the list. Actors are always in a rush, but it takes time and the right timing. Do not annoy your agent, but definitely call from time to time to ask if more pictures or résumés are needed (definitely not more than every other month). If six months or so go by and you haven't received an interview, and no more pictures or résumés have been requested, then maybe it is time to consider a change.

Try to set up a meeting with your agent before you make this decision. There is often tremendous wisdom he may be willing to share with you as you consider moving on to something or someone else. If you can't set up a meeting or at least a conversation on the phone, then you are definitely making the right decision. The big easy.

How

Hollywood, and "show business" in general, is a very small and closely knit world. Throughout the years, agents have moved on to very important positions: presidents of major studios, heads of production, heads of networks, and casting directors (to name a few). So when the time comes and you feel the agent you tried so hard to get to represent you isn't doing the job you hoped for, or doesn't have the kind of personality you prefer, or hasn't returned your "anything going on" phone calls expediently, be careful. Yes, it may be time to find another agent, but you should be professional and courteous doing this. There is a very good chance you will run into this person again in the future.

DO!

My suggestion is as follows: thank the agent for the time spent, acknowledge that you will be forever grateful for the opportunity you were given, and explain that you are sorry it didn't work out but you are certain it is time for you to explore other options.

DON'T!

You should not just write, "According to clause six of the Screen Actors Guild contract, you no longer have the right to represent me." Clause six has to do with how many days must pass before you can release your agent. If you are not SAG the clause does not apply to you in the first place, as you would not have signed a SAG contract. Also do not write that the agent is no longer your agent and that is that (or anything to that effect). You will never know how many times your agent tried to get you in for various roles or pitched to casting directors, and for whatever reason it didn't work. So be nice, and nice might just come back to you in the future.

Do not give your agent ultimatums such as, "If you don't get me something soon . . ." No agent worth your time will respond well to this. If you feel it is necessary to make a change, make that decision and stick with it. Share it with your agent and see what both of you can learn from your experience working together.

Special Note: I have been an agent and/or partner in an agency for more than thirty years, and believe it or not, I've seen agents get very angry when they were forgotten by a client at Christmas or some other important time. Maybe the client felt that the commissions they were bringing in were sufficient, or that they were getting their own work, or that the residuals were plenty, but clearly the agents did not feel this way. Remember, first and foremost, this is a business—take care of those who are in a position to take care of you.

At one agency I worked for many years ago, paying my dues, I saw this directly. When it came time, after pilot season, to clean house and let some clients go, those who did not bring in Christmas goodies (and were not booking jobs) were sent their release.

Even worse, at another agency, some of the agents got so mad when the boss got all the gifts, while they felt they did most of the work, that they did not submit the clients they felt had snubbed them. And the clients never knew! I do not agree with this kind of behavior, but I can understand their frustration. More importantly, *you* should be able to understand. Even though it is a business, these are human beings and they have feelings. All good business owners understand that about their clients and their suppliers.

1953. Working with the Bowery Boys was a great thrill. Many years later, I became Huntz Hall's agent.

1954. I was cast as a young punk, one of my first acting jobs. Chuck Connors was starring in this movie, and many years later I become his agent.

O. My first only starring in a feature. shot on loca- in Chillicothe, souri; the ctor was from e. I played a for the first last time.

47

A signed fan-club photo. This was my personal favorite photo throughout my acting career.

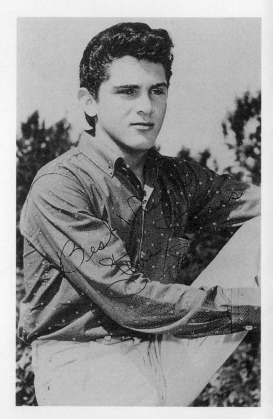

The Jack Benny Show (1960). *Working with Jack Benny was a real treat; that's him on the left, wearing a black wig.*

Left, At the Golden Globes with Annette Funicello, trying to be cool with a cigarette (1960). Below, with Roberta Shore on the set of "Annette" (1957).

Above, *Starring with Thomas Mitchell in Paramount's "Big Moment" (1950). This was a short film for some sort of charity. Mitchell starred in* Gone with the Wind *and* It's a Wonderful Life. Below, *guest-starring on the television series* Target the Corrupters, *which starred Stephen McNally (1961).*

Left, *Dressed as a cowboy circa 1951.* Below, guest-starring as a runaway prince in the early television series The Lone Wolf, starring Louis Hayward.

In 1957, Annette Funicello had her first starring role in her own series on the Mickey Mouse Club. It was called "Annette," and I had the pleasure of being one of her costars. Even back then, Annette was a very special lady.

With my pal Tim Considine on the set of "Annette" (1957). We have remained friends all these years. Tim went on to star in several Disney movies and on the television series My Three Sons.

Right, *On the set with Annette and Tim, 1957.* Below, *Working with the incredibly talented Betty Hutton was for me the most exciting time in my acting career. Miss Hutton starred in* The Greatest Show on Earth *and* Annie Get Your Gun.

CBS

BETTY HUTTON

Back on Television in

"THAT'S MY MOM"

Also Starring: *Steve Stevens*

HEAVENLY VIEW

modern screen®

TWO BUSY YOUNG STARS, SOARING
TO THE HOLLYWOOD HEIGHTS,
DISCOVER A FABULOUS
WAY TO TELESCOPE THEIR FUN
INTO A FEW PRECIOUS HOURS.

HOLLYWOOD is a dreamy place to live. Shelley Fabares has always known that—because Hollywood is where she's lived all of her 16 years. But she's viewing her fabulous city through even more excited eyes these days, since 19-year-old Steve Stevens became her special guy. In every favorite, familiar pastime, there's a new delight. Take a visit to the Griffith Observatory, for instance. The Observatory has always been one of Shelley's special places. The stellar lectures there are like a visit to outer space. And since the building itself is perched on one of Hollywood's highest peaks, the view of all Los Angeles, sprawled out for miles below, is simply breathtaking. The excitement of spending a few hours on this magnificent hilltop is nothing new to Shelley. But to while them away with someone extra-special, that is something new. And Shelley—and Steve—will have you know, it's simply heavenly! ●

The Hollywood publicity machine at work. Shelley and I were friends and I guess Modern Screen thought we would look good together for this makeup layout.

54

This layout was done before I worked with Annette. We had met a few times before and I guess someone thought we looked good together. I think they were right.

Charmed by Steve over a sundae, Annette was later disenchanted when he agreed with her mom she was too young to own car.

PDC

TV PICTURE Life

OCTOBER • 25¢

In the midst of great fun, Steve will remind Annette it's time to go home. Annette will turn sweet sixteen in October.

• If Annette Funicello ever wondered where she stood with Steve Stevens, she found out somewhere around her second or third date with him. In his eagerness to pick her up, Steve drove so speedily he piled up his car en route— but calmly proceeded to keep the date anyway! Still another Steve (an 11th grader by the name of Elkins) is also showing Annette the thrill of a first romance. Almost philosophically, she refuses to choose between them. "What will happen will happen," says she, "and for the best."

TV Picture Life

On the set of Star Trek VII with my client and friend James Doohan. James has bee[n] a client for twenty-eight years.

With George C. Scott on the set of Rage (1970). I was the casting director and Mr. Scott was the director. This was the beginning of a very special working relationship.

On the set of Oklahoma Crude with George C. Scott (1973).

Clowning with my client and dear friend Chuck Connors. He was a lot more than just "The Rifleman."

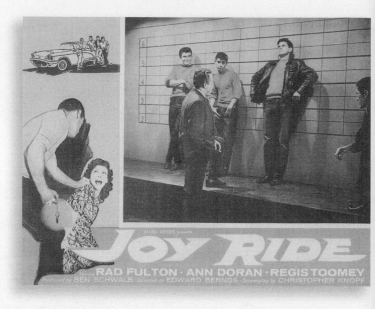

Cast again as a punk in 1958. I thought I was better at comedy.

Guest-starring with Roy Rogers and Dale Evans (1958) was a fantasy come true. We kept in touch through the years, and their daughter Cheryl and I are still in contact.

Producing a Movie Was a Dream

Nightmare on Elm Street

I am no different from any of you. I came to the industry with quite a few dreams and fantasies. From my first break as a young actor, more than fifty years ago, I dreamed of producing one of those magnificent films that begins after the familiar and grand music accompanying the 20th Century Fox or MGM logo.

Well, I got my chance in 1997. A well-established producer approached me to form a management company and produce films with him. I jumped right into that pond, without giving a second thought as to whether or not I could swim. The allure of being a Hollywood producer was too much and more or less overwhelmed my common sense.

So, suddenly I had to do something I never thought I would do: become a manager. Why? Because you can't be an agent and a producer at the same time (no such rules apply to managers). So I took most of my longtime clients with me and started a management company. I was able to get most of them parts in the two films I produced. But little did I know there would be an ambush just around the corner.

The ensuing melee turned out to be good for many, but most certainly not for me. My oldest son, Mark, a schoolteacher, wrote one of the scripts, so he was able to get a feature film in the theaters. My youngest son, Steve Jr., was first AD on both films and watched my back as best he could. He got an unparalleled opportunity to learn what *not* to do as a director and producer, as it turned out. Clients like James (Scotty) Doohan, Bernie Kopell (*The Love Boat*), Ty O'Neil (*Mighty Ducks 2* and 3), and Anne Lockhart were cast in one of the films, playing roles they would not otherwise have had the opportunity to portray. So as I said, it worked out great for some people, but for me . . . it was a whole different story.

The person I was working with had an ego the size of Las Vegas and the glitter to match. He would consistently hire good, professional people and frustrate them when they tried to do their jobs. I spent most of my time convincing Teamsters not to beat him up. He had no clue how the little guy depended on a day's wages to support the family. He seemed to go out of his way to take his sweet time in sending payments to extras and day players. I could not understand his behavior at the time, and I still don't.

He had his own idea of what he wanted to put on film, right or wrong, and would not take anyone's input. So the two films turned out to be like by-the-numbers paintings, and just barely good enough to get releases. It was the most stressful time in my professional career, and I wound up in the ER with a stress condition that the doctors thought might be a heart attack. Lying on a gurney, looking up at several strangers talking about things I could not comprehend, and fearing I would soon be dead made me realize that producing is for young, fearless people who are stronger than I am. I do not regret taking the shot that was offered to me, because if you don't take chances you could miss out on your dreams and lose sight of your goals.

But don't let my story discourage you. The important thing to take away from it is that your dreams may change as you change; your engagement with the industry will shape and mold who you are and who you become. You will grow and discover other dreams that you couldn't have imagined when you began. It may turn out, much as it did for me, that you didn't fully understand your dreams when you laid the groundwork. That's why it's important to bring your entire presence to whatever aspect of the industry you pursue. Whether you are producing, acting, directing, casting, fetching things, or interning in craft services, bring the full complement of your talents to bear. There is no job too small, no task too demeaning, nothing you can do in any given opportunity but explore the depth of your talent, commitment, dedication, intelligence, communication skills, and understanding, as you simultaneously strive to improve them. If you do this, there is no dream you will fail to explore.

Do I Need a Manager?

And Other Questions You're Afraid to Ask

Just as with agents, there are good managers and there are bad managers. If you can't find an agent that will represent you, then a good manager might be very helpful. An agent or agency will have anywhere from sixty to several hundred clients. Usually a manager will have just a handful.

Unlike agents, who have to renew their state license yearly, managers are not regulated by any governing body whatsoever. They can choose to be licensed by the state of California, but they are not required to be. Some managers take it upon themselves to be licensed by the state. This can give you a little insight into the type of manager with whom you'll be dealing. The qualifications for a state license are that you have never been convicted of a felony, hold a $10,000 bond, and maintain an office not attached to your home or a nightclub.

A manager does not have to abide by the 10-percent commission rule that agents do. Most take 15 percent, but over the years I have heard of some taking much more. That is one of my main gripes with managers: They do not have to maintain an office, legally they cannot find employment for a client, and yet they take a bigger percentage than an agent.

If you have an agent, discuss the possibility of a manager before you go out and sign with one. Some managers want to be the Chief and reduce everyone else to Indian. It should be a partnership, with everyone working to further the actor's career. Unfortunately it does not happen like that all the time. Managers are only as good as their background and showbiz connections. I have found that the best managers were agents at one time and have an understanding of the business. They know what an agent goes through day to day. They also understand the reality of today's marketplace. Many managers, like some agents, have no showbiz background. They might be a former actor, or even a family member of the actor. Frankly, many managers just decide one day to become one. *Poof!* "I'm a manager."

There are many horror stories I could relate, but I will limit myself to three. I am sure managers have *agent* horror stories as well, but I'm writing this book so I get to tell mine. If they want to tell theirs, they can write their own book.

- I booked an African-American actor on the feature film *Maverick*, starring Mel Gibson. I was very lucky. I had booked eight actors in good roles on this film, and all worked for several weeks. This particular client was a respected actor, but he had not worked for several years. He picked up a young (very young) African-American manager who believed his new client was still a big star. Now, there were so many actors on this production that only really big stars got "star" dressing rooms. All the others had their own honey wagons. This young manager showed up on the set and demanded the star treatment for his new client. He even played the "race card," to the shock of everyone. He succeeded in getting a star dressing room for his client, for all of the rest of that day. At the end of the day, he was informed that his client was no longer needed. All the other actors in that same scene worked two to three weeks. So this young manager more or less cost his client two weeks of work.

- Just a few months ago we met a young actress who sent us a picture and résumé. She came with a manager: a nice lady, but one with no real managing background. We set up a date for the actress to come do a scene. She was great! We said "let's do it" and "get us some pictures right away." The manager said that she would take care of the pictures the next day. Three months went by and we never got any pictures. The actress called and left a message on our voicemail wanting to know if the pictures were working and if there was anything she could do to help. She didn't leave a number, so I couldn't tell her that her manager never brought us the pictures. Another month passed before the actress called again, and she was quite surprised when I told her that when we never got any pictures, we assumed she had gone with another agent. She was particularly shocked as the manager, who was out of town for three months, told her she had taken care of everything. I get the chills just relating this story.

- I had one client that I not only got started in his bid to star in a feature film, but whom I personally helped support. I even got a producer to write an episode of *The White Shadow* to show off his unique talents. I was in the middle of making a deal for his starring role when I got a call from a manager who informed me that my contract with his new client had lapsed the day before, and his new client was no longer represented by me. He would be moving his client to a bigger agency. There was an option for a sequel attached to this film, and they wound up making five plus a short-lived TV series. Though I do blame my former client for this inconsiderate behavior, the manager did promise him the world and a bigger agency to represent him. So . . . yes, I am a little gun-shy of managers who come out of nowhere and guide the life and career of a novice actor who is depending on them for good advice—not only in how to be a star but how to be a positive part of this industry.

• • •

Do you need a manager? The more good, responsible, knowledgeable, and ethical people you can get on your team the better. It's a win-win situation. Just be careful and do your homework.

Stealing Roles

*T*here are so many ways you will learn to steal roles between now and when you look back on your career and try to figure out just how you "did it." When I say "steal a role," of course, I mean exercising all your skill, knowledge, connections, and luck of the draw to tip the playing field to your advantage. It's more like stealing second base than stealing a loaf of bread, but sometimes it's like walking into a saloon with both guns drawn. You'll see people stealing roles all the time, and sometimes you'll lament their advantage and sometimes you'll admire their chutzpah. And sometimes it will be you stealing a role, and you'll feel proud and wise, and you'll probably make a footnote for your memoirs. It's all part of the game, there's just no denying it.

Wardrobe Makes the Actor!

Dennis Fimple was the consummate professional actor, as well as a client and friend for over thirty years. Sadly, Dennis passed away last year. I often find myself thinking of this wonderful character actor and the important lesson he taught me, which I would now like to pass on to you. Dennis lived in Frazier Park, which is about sixty miles north of Los Angeles, high in the mountains. He had a van that contained as many wardrobe changes as a costume store. Early on, he found that too often he would be in town on an audition for a dirt farmer, and I would call him and say he had another audition for a nerdy professor later that same day. After only a few times driving all the way home to change and then back to town to audition, he landed on the van/wardrobe shop solution. Dennis was a chameleon and could play so many characters. He felt the wardrobe helped him steal the roles, and he was right.

He would search high and low for the wardrobe that would be just perfect for him, and many times the production would wind up paying him extra to wear his own clothes on the film.

I suggest you learn from this resourceful actor. You may not need to drive around L.A. in a van/wardrobe shop like Dennis, but it's a good idea to have a small "audition" wardrobe in your trunk. Whatever roles your agent typically sends you out for, you should be able to dress that part in a heartbeat. Now I'm not suggesting that you show up for a cowboy role in chaps and spurs—though many do, and believe me the closest these people have been to a horse is the droppings at the Rose Bowl parade. This is one of the things that has not changed since I was acting forty years ago. Wardrobe helps get you into character, and shows the director and casting director that you can comfortably inhabit whatever concept they have selected.

In theatrical productions you can reduce the look to more of a suggestion, but you should go as far as you need to—as far as you find beneficial. A sport shirt and slacks is not going to work if you're auditioning for a farmer, just as jeans and a T-shirt is not going to do you any favors if you're hoping to portray an upscale lawyer. For the ladies, whether it's a character or just "beautiful/sexy," you should expect to have an audition wardrobe that you don't wear anywhere else, and be prepared to do punk, bag lady, soccer mom, trophy wife, sexy lawyer, hooker, or prim-and-proper prude at the drop of a hat, scarf, or Gucci bag.

You will hear all kinds of different opinions on this subject, but I have seen it work too many times as an actor, a casting director, and an agent. It's exactly the same as my advice in the chapter on photos, and you want to approach it the same way. Ask yourself: *How can I use every single thing in my arsenal of talent, makeup, wardrobe, power, and luck to steal a role away from the competition?*

Spend a few days searching the many thrift shops in your area, and you will be amazed at what you will find. There is a unique store called "It's a Wrap" In Burbank that sells wardrobe from feature films and television shows at respectable prices. So load up your trunk with several changes of clothes and you will never be caught without just the right thing for the next audition that pops up like a rattler on the side of the trail.

Makeup Makes the Actor!

Remember, every advantage you have, every skill you mas-
ter, every extra talent you were born with or have struggled
to embrace will carry you a little closer to your goal and
just may be the infinitesimal advantage that scores you the
big break. I cannot tell you as much about makeup as I can
other areas, but I happen to have a best friend from my
days on the *Mickey Mouse Club* who won an Emmy for
makeup, and that's good enough for me. May I introduce
my longtime friend and Emmy-winning makeup wizard,
Tommy Cole.

My good friend from the Mickey Mouse Club, *Tommy Cole.*

Advice from the Makeup Room

Sixteen years as an actor and thirty-seven years as a makeup artist have made for a great "ride," one filled with invaluable learning experiences, good times, challenges, and opportunities. Through these years in TV and film, I've followed certain rules—tenets if you may—ways to conduct myself in my life both professionally and privately. Not that I'm all-knowing, for I'm not! Like all of us, I've made mistakes, but I've hopefully learned from these errors and have become better at my job because of them. What I'm going to impart to you are my thoughts and ideas from experiences along the way. If some advice works for you and helps, great. If not, well, that's OK too.

Be on Time

Being late for an interview can be the difference between you and someone else—even someone less talented—getting the role. It's that important. Being late for your call might get you fired from the job, but, even if it doesn't, it will raise the ire and disdain of the director and ADs. All of us, at one time or another, will be late because of something beyond our control. So, as a rule, make a habit of being early and leave enough time for the unexpected. You'll not only gain the respect of your peers, but you'll gain a reputation for being reliable.

Know Your Priorities

As a makeup artist, the first people I made friends with were the Director of Photography and the Transportation Captain, for the DP enhanced my artistry and Transportation made my life so much easier logistically. As an actor, the first people I made friends with were the DP, makeup artist, and hair stylist.

These were the most important people to have on my side, for they were responsible for the way I looked in front of the camera.

Makeup Hints for the Actor

Know your face. Practice the basics of makeup so that you could do your own makeup in a pinch, or just in case you get someone who applies cosmetics like a house painter instead of an artist. Also, by knowing what works for your "look," you can give input to the "puff and fluff" department as to your character. Usually they will be very accommodating, unless they have special instructions from production, or if it's a period piece. If it is period, let the artist do his or her work. And don't add to or change it in the dressing room. You don't want to get yelled at by the makeup person or the director for not looking the part. If asked, let the artist do makeup on your hands, so that if you're going to have your hands near your face, your face and hands will match. There's nothing more distracting to the eye than an actor who has a natural, made-up face but hands that look like they are from the morgue.

Now to Your Hair

Whether male or female, come to the set with clean hair. It's difficult for a hairstylist to do something creative if your hair is dirty. If you feel strongly about a certain "look," comb or brush your hair the way you like it and have the stylist fix anything sticking out, spray if needed, and say, "thank you very much." If you're working on a period piece, be prepared to have your hair cut and styled for the vintage look of the time. If you won't let them cut your hair, you shouldn't have taken the part. If female, it's helpful to come in rollers. If you can't, be prepared to spend a longer time in the chair.

Be Prepared

Sounds like the Boy Scouts, I know, but know your lines. Acting is like anything else in life, the more you practice and prepare, the better you are going to be at your craft, thus gaining the respect of your peers and the people who hire you. There's nothing more embarrassing than to be with other actors who have done their homework, while you keep asking, "line please!"

Be Aware of Hygiene

That includes body, breath odor, and cleanliness. Check your nails every now and again. You also might think twice about having that bowl of pasta full of garlic, or that delicious hamburger laden with onions, or that quick smoke just before doing a job. If you must indulge, please brush, floss, gargle, spray . . . do something! Hopefully your friends will let you know if you might be a little bit offensive.

Lastly

Treat people the same way you would like to be treated, with fairness and respect! Don't gossip or backstab! Be generous, unselfish, and sharing amongst your peers. As an actor, you are in a wonderfully fulfilling profession, one that can be very rewarding artistically and financially to those who are willing to work hard at their craft. Take pride in what you do and never stop learning . . . never!

Good luck, and have a wonderful and exciting career.
—Tommy Cole

• • •

Listen to Tommy, the man knows what he's talking about and has given you more than a few ways to steal roles. I'll leave you with one more. When reading for a role, try to find something in the lines or even in the persona of the character that would be interesting to the casting director, director, or producers. In other words, find the humor in the drama or the drama and "edge" in the humor. Find a little quirk that might work to make the character memorable or even find a completely bizarre quality (without going overboard). Give them something that will make them remember you after they've read all the other actors. More than once when I was casting, the director would say, "Get that guy who did that thing, I liked that he took a chance." The director will usually prefer to bring a capable actor "down a notch" than to try to elicit a performance from someone who isn't ready to give it. It's much easier to hone what is there than to pull out what isn't. Go for it! Wouldn't you rather feel the nervous energy of having taken a shot and given your best than be stuck in traffic thinking what if you'd done this or that? "It would have been really great if I'd . . ." Too late.

Be prepared to steal. Go in with the idea of just what you're up against, and how you've tipped the playing field to your advantage. Have a plan to make yourself memorable, pull out every stop no matter how trivial, and execute your vision. There's a great old saying that goes something like, "If you don't know who the mark is, it's probably you." Well, if you don't know how you're going to steal a role, it's probably about to be stolen from you. Have fun with this—piracy has a long tradition of enjoying a good sense of humor.

What Has SAG Done for You Lately?

The Screen Actors Guild cannot and will not do anything for you if you are not a member. I have an honorable withdrawal from SAG because you can't be a member and an agent at the same time. So, in reality, I am still a union man. I was brought up believing in unions; my dad worked on the docks for many years, and he sure believed in *his* union! He later became a member of SAG, and he was so proud of his membership card you wouldn't believe it. I personally believe that SAG today is not the powerhouse it was fifteen or even ten years ago. Its membership is so divided on so many issues and spends so much time and money trying to resolve them that it has very little time for the day-to-day actor trying to support a family.

SAG Vouchers

SAG vouchers are like gold for the beginning actor who wants desperately to become a member of the guild. Having that card makes it easier to get representation, and to be seen by casting directors for commercial and

With Steve Jr. on strike, 1980.

episodic work. If you receive three of these from working as an extra, you can join SAG as a full member.

But be careful. I have heard horror stories about second ADs and even third ADs doling out these vouchers in exchange for sex. I am not kidding. Recently I have met several young actresses who have told me they were confronted with this kind of exploitation. I do not know to what degree the SAG board is aware of this, but it's that same old bullshit. Someone with a little power, like an assistant casting director getting paid to teach an acting class, comes up with a different slant on the casting couch. Whether it's money or sex being extorted from fresh young actors trying to parlay a little talent and a lot of guts and determination into a life in the industry, it takes a particular breed of scoundrel to exploit that desire for profit.

Financial Core

SAG does not want you to read the following information, so if you are in the business to please them, go ahead and skip directly to the next chapter. Now, what *I* want you to know is that whether or not you are currently a member of SAG, *you* must ultimately control your destiny, not the union. No matter what you've heard, it *is* possible for an actor to join the union but still be able to work on non-union projects, sign with a non-franchised agent, and even work for whatever form of payment he or she might choose. Yes, you can even work below union minimums. One way is to be a producer on the project and therefore be working for yourself, and another way is to declare yourself "Financial Core."

Financial Core status is your fundamental right as a

member of SAG, whether you work in a "union" state or a "right to work" state. Financial Core enables you to remain a member of the union in good standing while paying significantly lower dues. The only things you will miss out on are the "bonus" privileges, such as movie screenings or voting for awards (though you can still be nominated for awards). Most people will not even know you have declared this status. You might wonder if there any big-name actors who have done this. Believe me, there are many.

Here's how Financial Core works. The Supreme Court passed down a landmark decision in the case of *Communications Workers of America v. Beck* that has been hailed as a victory against compulsory unionism. It is meant to reduce the power that the union holds over its members, production companies, and non-union actors.

If you are a SAG member (you must first become a member of the union for any of this to apply) and you wish to declare Financial Core, all you need to do is fill out a form supplied by SAG. The form basically states that you wish to resign your "full membership" and maintain "Financial Core" status. Return it along with your union card, and presto! Financial Core members are not allowed to vote in union affairs and are not eligible for other fringe benefits like the newsletter. On the other hand, you may pay reduced dues, since none of your contributions will go to non–wage related activities. The most important benefit is that you will no longer be restricted by the traditional SAG work agreements. That is, you can work how, where, and for however much you want. You can work below SAG minimums or you can negotiate a much higher payment or even deferred payment. You will remain a member of SAG and thus be eligible for all SAG-related work (no one needs to know your status), and you can act in non-union films, commercials, and television. SAG will represent you in wage-related matters only, and then only when it's a SAG production. You will be your own representation in non-union material.

If you decide you want to declare Financial Core, send a certified letter to the Screen Actors Guild Legal Department (5757 Wilshire Boulevard/Los Angeles, 90036) expressing your wish to "declare [your] Financial Core status." They may try to talk you out of doing this. Here's a quick breakdown of the pros and cons to help you make your decision:

PRO	CON
Take any work (potentially work more)	Cannot run for SAG office
Negotiate own pay in non-union work	Cannot vote in union business
Pay lower dues	No newsletter
Maintain health/ welfare benefits	Cannot belong to the film society

It's a decision only you can make. Do you want to take non-union work to increase your chances of getting spotted, or do you want to make sure you have a vote in the next SAG referendum on merging with AFTRA? Of course, the power to negotiate your own rates on non-union work may be a Catch-22 in that most non-union films don't have money to pay you even if they want to. But it does represent a big opportunity to do more work and to potentially get noticed. Many well-known actors have done it, and ex-SAG President Charlton Heston fully endorses it.

If you want to reinstate your full SAG membership at a later date, you will be allowed full membership solely at the discretion of the SAG board. So it might not be something you can easily undo. However, it is your right to work as an actor; you should not be made to feel as though you are breaking the law to ply the trade you so love.

Taft-Hartley

The big thing to remember about Taft-Hartley is that it is not an entertainment-based law. It's a rule put on the books that applies to all unions. It provides for the fact that everyone can be a part of a union; they need only take the work. Within the entertainment industry, producers are saddled with a small fee if they choose to work with non-union actors. The actors are given the work and the capacity to join the union in the next thirty days. The producer has the choice of whether or not to hire the actor, but they are required to acknowledge the status of each actor to the union. What this means to you is that you should audition for whatever moves you, and the rest should take care of itself.

SAG can be both your friend and your enemy. Until you have joined the ranks, the union will bar your way to many auditions. Once you have, the union will try to protect you through many trials. The conflict will be how you perceive the union—is it serving your needs? Today the divide is so grand, the union may lose its importance across the board. Don't let SAG define you. Define SAG the best you can. The more people there are truly caring for their fellow man, the better. Don't forget that DPs, gaffers, and best boys have as much to do with the industry as any actor. If you don't find that the laws are helping you, check them out.

Theater in L.A.

Give My Regards to Santa Monica Boulevard

There are more theaters in the greater Los Angeles area than in any other city in the United States. The opportunities this presents are myriad, though perhaps challenging. Theater in L.A. is completely overshadowed by the film and TV industries. That's where the money is, and in this sunny coastal environment, theater just doesn't stand a chance.

The primary danger this creates is that many actors don't take it seriously. They treat theater as a sort of grand and complicated audition piece. Their focus is on which industry people will be in the audience that night and not on the complexities and beauty of the specific piece of art they are about to share with the world. This makes for some spectacularly bad theater.

In New York City, on the other hand, the theater is

such a well-established tradition and tourist magnet that it doesn't take a backseat to other artistic or entertainment-industry endeavors. This leads to a kind of competition and seriousness that creates some terrific works of art, although it can stifle the fun and inventive side of theater. In L.A., if you look carefully, you will find some groups that are doing the work for the work's sake, and are not at all concerned with the big boys in TV and film. This detachment from the larger commercial forces creates an opportunity for some daring and innovative work to be done.

If you can find the right places, and more importantly avoid the wrong ones, you can really get involved with some important and groundbreaking work. You will be exposed to tremendously talented people who are pursuing this for the love of artistic expression. You might be able to learn more from these people than from all the members of the "capital I—Industry" you'll eventually encounter.

Now, you might be saying, "But, Steve, I really *do* want to be seen by agents and managers and 'the Industry.' What do I do?" The key is to focus on the project at hand. First of all, you can't completely avoid theater projects that seem to be about "getting discovered." It is very difficult to weed out the people who will ultimately have that as their only goal. The more you can avoid those kinds of projects the better, but actors have to act. Where there is an opportunity to perform you must seize it.

So keep your eye on the ball. Who attends the performance is ultimately of little or no consequence. That you may be in the midst of getting your big break is completely irrelevant to the moment of being onstage and expressing your artistic talent. Let those concerns be what they are—secondary. They will happen or not happen regardless of your concern with them. If you truly think you may give a better performance if so-and-so is there, you should really reevaluate your motivations. Take the art seriously, and serve the story as well as you can. Focus on that and your performance will be that much better. Industry observers can usually spot grandstanders, and it's a turn-off because it's such a selfish act. They'll have a good idea of the type of person you'll be to work with even before they assess your acting skills.

There are a number of small theater groups that function on a membership model. These are usually nonprofits operating under an Equity "under-99" contract. If you have joined SAG, AFTRA, or Actor's Equity, there are restrictions on what types of productions you can be a part of and how you must be treated and paid. The unions are pretty cooperative about giving waivers for these small "under-99" contracts (the name derives from the requirement that the theater have fewer than ninety-nine seats). You will usually pay the theater group a monthly membership fee and commit to volunteering a few hours a month. There's a pretty wide range, but typically you should expect to pay twenty-five to fifty dollars a month and volunteer about ten to fifteen hours of (non-acting) work. This would consist of running the box office, concessions, lighting (if you have that talent), or building sets and maintaining the space (cleaning up, etc.).

Some of these theater groups have spaces of their own and some do not. Usually there will be between fifty and one hundred members, with probably thirty to forty "active" members, and most of the work being done by ten to twenty people. If you consider a season of eight six-week runs and an average cast size of about six or so, and an average "lead actor" size of two, you begin to see the difficulty. Often it becomes a highly political process getting good parts in these groups. However, joining one will give you the opportunity to appear on stage (most promise at least two roles a year—without size guarantees), and much more importantly it will give you an opportunity to learn. This is a microcosm of the industry at large; the skills that will help you become a leading, continually working actor in a small theater group are the same skills that will keep you working throughout your career. Talent, dedication, hard work, and perseverance are necessary, but a little luck, training, and trial-by-fire will help too.

I suggest you frequently attend plays at these small theaters. There really is a great deal to learn from the patient and dedicated study of your craft. Watching the mistakes or successes of your fellow actors, in an intimate space where you have a good view of what they are doing, can be very educational. Not only might you learn some acting nuances, you will discover which of these groups is doing work that appeals to you. As I said, getting in with the right group can make all the difference. These people will become your friends, peers, teachers, and, sometimes, family. As they make it in the industry, they will remember the kind of actor you were and whether your focus was on being discovered or helping create great works of art— which, in turn, helps everyone get discovered and maybe even makes the world a better place.

You should also consider finding the right play and

working on it with some friends as an alternative to doing showcases. On one hand the showcase has taken care of some of the work for you, in that industry presence has been arranged. However, showcase acting is almost always a step below average, and agents and casting directors (while trying to spot talent) will really end up focusing on types and broad physical characteristics. If you have a group of friends that you think you might enjoy working with, it's quite easy to rent one of these "under-99" theaters. They run between $1,000 and $3,000 a week, depending mostly on size and location. If you consider a three-week run at a bargain space, you should be able to mount a play for just under $5,000. If you can come up with a play that has four or five good parts, it gets the cost per player down to around the size of an expensive showcase. When you factor in that a showcase will typically be just one or two performances at about 8 minutes of stage time each, while the play will offer about twelve performances at 90 minutes (even if your part has only 40 minutes of stage time, that's 480 minutes on stage), you can quickly see how the idea of doing a play is far more valuable when measured by dollar per minute of exposure.

Of course, you will have to work to get the right industry people to attend. Some industry types will simply not attend at all, end of story. However, if you put on a really good play, and entertain, teach, enlighten, and bring out all the wonders that come from seeing a truly brilliant piece of theatrical art, I will guarantee you that there are good people in this industry who will take notice. And they might just remember how you touched them during that play for the rest of their lives. I doubt any showcase could offer to do the same. Create really good art . . . it's a sure-fire strategy for success.

Santa Monica Boulevard stretches 14.26 miles from the ocean in Santa Monica to its intersection with Sunset Boulevard in Silver Lake. Along the way it boasts quite a few of these little theaters. It's not Broadway, but on or near this main theatrical thoroughfare you will find a unique collection of theater groups doing art for all the right and wrong reasons. Couple that with all the little the-aters dotting the greater-L.A. cityscape from Hollywood to the beach communities to Orange County, Pasadena, Silver Lake, and the budding NoHo Arts district, and you've got places to work, people to meet, and art to share with the world.

Beyond Hollywood

It's Not Just for Actors Anymore

*T*his chapter is for anyone who does not live in Los Angeles and does not want to live in Los Angeles. Are there any of you out there? Raise your hands.

Vancouver's sex-trade workers (streetwalkers) must be compensated for displacement they experience at the hands of a production company if it occurs during their scheduled operating hours. The same holds true for homeless people who are pushed from beneath a bridge or doorway or asked to leave a park. Financial compensation must be provided for all disrupted work, including panhandling! The *Vancouver Sun* supports this campaign, saying: "We see no reason why any unorthodox entrepreneur should be treated differently from other businesses when it comes to compensation" (quoted in *Variety*, August 21, 2002).

When a movie production goes to a location away from Hollywood, millions of dollars can be injected into that community. Many times, the same people reap the benefit of those dollars over and over. Here are a few tips on how you can get your own taste of the Hollywood pie.

If you own a business, any kind of business, and a movie production films in or around the town where you live, you can become a part of that movie production. Would you fit in as an extra, an actor, a stand-in, or a janitor? Of course you would. Do you own a catering business, a restaurant, a dry-cleaning business, a car-rental business? These and many more services will be needed when a Hollywood production comes to your town.

Most states have a film commission—paid for by your tax dollars. A production company contacts that commission to find out what locations might be available that fit their story. They will also check out what businesses are in the area to support their needs. They'll angle for quite a few freebies, such as no-fee permits for the locations and the cooperation of the local police force. There is quite a list of details that might make a difference between one location and another. There's also a lot of money at stake and many hands outstretched. If you play your cards right, however, you can have a wonderfully rewarding experience helping the Hollywood magic find its way to the silver screen.

DO!

When you hear the rumors that a movie is going to be shot in your town, chances are the deals have already been struck. You will be left standing behind the yellow tape trying to get a glimpse of Jack Nicholson. However, you *can* call your film commission and request a meeting. Don't take no for an answer. Remember, the film commission is paid for with your tax dollars. The head of each film commission is usually appointed by the governor. If you have a business you feel could be utilized by the film company, prepare a résumé for yourself and your business. There's no telling who you'll actually get a meeting with, but tell them you want the opportunity to present yourself and your business to the production manager.

If you are a local actor in the hometown theater group, find out from the commission the name and contact information for the local casting director. Most big cities have local talent agencies, which are sometimes combined with modeling or acting schools. (In Los Angeles that is a no-no for agents.) The local agency should know who is doing the local casting and will get first shot for their clients. You know the drill: big fish, little pond. However, a good casting director (local or otherwise) should want to meet all the talent available.

Unfortunately, this is not always the case. I experienced this problem firsthand when I was a casting director. Instead of using the local casting director on several films, I went to the location and did the casting myself. I contacted several of the local agencies in advance and had them send me pictures of their clients. When I got to the location, I met all of them. I also received many pictures and résumés from local people who wanted to try out for the film and were being stopped by the agencies and a few local casting directors working with the agencies. The agencies were bombarding me with their people in an attempt to keep me from meeting with actors who were not in the loop or did not know how to play the game.

My job was to see everyone for my director and spot those rare and wonderful faces you seem to find only in small towns. I even went to the unemployment office in Stockton, California, while casting *Oklahoma Crude* (the George C. Scott film) to troll for undiscovered talent. I sat there all day and eventually picked twelve "types" to work on the film for the next several weeks. I also saw several actors from San Francisco who came down on their own and landed nice parts in the film.

Let me explain the reason a production company wants to use as many local actors as possible. It's strictly logistics—dollars and cents. One of the reasons a production leaves Hollywood is to save money. Hiring a local actor saves the production in many ways, including: transportation (which must be first class), lodging (high standards), and a per diem (calculated weekly). Additionally, they do not have to pay a local actor intervening days (which means you don't get paid for days you don't work). So, for the local actor, business person, or fan just looking to help, how do you make the system work for you?

DO!

You can find out what studio is doing the film by calling the local newspaper. They probably already know and are planning a story on it. You can also try the local chamber of commerce. If that doesn't work, call the film-commission office. If you still need more information, get a group of friends together and split the cost of a subscription to *Variety*. It lists all current and future productions complete with the phone number of the production company (www.variety.com; for subscriptions call 1-866-MY VARIETY).

Call the production company to find out the contact information for the main casting director and send your picture and résumé. Let them know that you live near where they are planning to shoot the film and you would like the opportunity to audition. If you are a good type for the film, the casting director (CD) will meet you when the pre-production team gets on location. The CD may also instruct the local casting director to meet with you. If that doesn't work, send another picture and résumé to the producer or director when they get to town. It shouldn't be too hard to find out where the production office is. In other words, do not give up. If the story is a period piece, get a friend to take some pictures that make you look suited to that time and place. If it's a Western or upscale city location, do your best to look as though you fit the surroundings. This will give you a jump on any actors just sending in a picture.

If by chance you have ever worked in a commercial or other local production and joined the Screen Actors Guild, you must be seen by the local casting director or the Hollywood casting director when they get to town. This is a SAG rule.

DO!

For businesses, find out what studio is producing the film (call the film-commission office for that information) and call the studio to request the production office for that particular production. Ask to speak to the production manager and offer to send a brochure on your business. It is his or her job to get the best deals for the production and save as much money as possible. Send the brochure via overnight delivery so that it must be signed for and gets there the next day. Follow up with another phone call in a few days.

If you have a great-looking farm or ranch, general store, or other unique location, take some pictures and make the same call to the studio. Ask for the production office of that film and ask to speak to the location manager. There is good money in location rentals. The $500–$5,000 per day that a typical location rental commands can make a few mortgage payments, not to mention provide some stories to embellish for the grandchildren about the time Brad Pitt was in your store, barn, field, or house.

If the location manager is interested, be flexible with your price. Don't think you can gouge just because this is Hollywood money. They are out to save as much as they can, and if they can't get the best deal at your place then they'll try someone else's. At the same time, make them pay the going rate and remember: everything is negotiable. Of course, if they need your racetrack and it's the only one around . . . hell, give them the location for free if they can find a small part in the film for you, your in-laws, and of course, the kids.

DON'T!

Don't wait until the last minute to take care of any of the above. When a movie production comes to town, everybody wants a piece of the action.

DON'T!

If you are an actor, don't wait until you hear a movie is being shot in your town. Try to get an agent, and always have your photo and résumé ready to send out. If you live less than a hundred or so miles from a big city, research the possibility of an agent there. Make yourself available for interviews if a commercial comes to that city. Get as much information as you can from the local Screen Actors Guild office. You do not have to be a member to get that information.

DO!

Try to get in good with the assistant director; he has more pull than you might guess. Try homemade cookies.

DON'T!

Do not try to buddy up with the movie stars. If they come to you, then it is OK. Some movie stars are like you and me, while others have big, big egos. They are there to work and then get back home to their families. You can use your judgment on which actors are approachable and which are not. It's usually pretty apparent.

I have seen actors and extras get kicked off a production because they asked for an autograph or "Can we get a picture together?" The two biggest movie stars that I personally noted to be very approachable were Mel Gibson (on the set of *Maverick*) and an altogether different kind of actor: George C. Scott. So give the stars their space, do your job, enjoy yourself, and go home.

Look at the Breakdowns

Go Directly to Jail

*N*o, this chapter is not about car trouble. A breakdown, in casting terms, is a document—print or electronic—that lists every role in a film along with descriptions of the characters. (See Chapter 16 for a sample breakdown.) Generally agents receive them, but sometimes actors want to see the breakdowns themselves. This is a practice the industry frowns on, and to explain why, I've gone right to the source. Gary Marsh is the president of Breakdown Services, and here he is to break it down for you directly.

Fact, Fiction, and the Future

Breakdown Services provides casting information to talent representatives. The breakdowns are created by staff writers at Breakdown Services who read scripts and create an objective synopsis of each character. The casting director approves the individual breakdown, and it is then released to talent representatives. Breakdown Services releases about two hundred different projects out of Los Angeles every week; a lesser number are released out of New York, Vancouver, Toronto, London, and Sydney. The breakdowns include television projects, feature films, commercials, and theater projects.

Breakdowns are generally restricted to talent representatives for many reasons. Of course, the question I am most often asked is: Why can't actors get the breakdowns? Certainly, if I were able to provide breakdowns to actors, why wouldn't I do so? Historically studios have restricted access to scripts. Scripts have never been generally available to the acting community. Actors have always wanted access in the belief that this information would help them get acting jobs. But given the limited time devoted to casting, especially in television, casting directors have found that if actors have access to what is being cast, the casting directors do not have time to do their job. They wouldn't even have time to wade through the sheer volume of actors submitting their pictures and résumés, let alone take care of the interviews, readings, and negotiations. Add to this the pressure imposed on casting directors to find top-name actors to play the available roles to further enhance the production, and you see that their jobs cannot be accomplished if they are besieged by actors who are not yet known.

Breakdowns help casting directors communicate their casting needs to talent agents, who then submit actors for available roles. If the system works, casting directors save time since they don't need to communicate directly with every agent and manager. So breakdowns end up being restricted to talent representatives who will hopefully submit union actors for union roles and submit actors for star-name roles who are actually star-name actors. There is hope, however, for actors.

There are projects that need and want direct submissions from actors. The majority of these films pay little, if anything, but for the beginning actor, these non-union, low-budget films, music videos, student films, SAG limited exhibition films, and industrials can provide not only experience but the material to create a video reel that can become an important promotional tool. With the advent of the Internet, Breakdown Services created Actors Access on its Web site at www.breakdownservices.com, which provides a way for actors to submit themselves directly for projects posted to the site. Actors can view these projects at no charge and submit themselves for roles. This is a free service because it is recognized that actors typically won't be paid or will be paid very little for acting in these types of projects. Acting in the listed projects, however, will give the actor experience and a chance to learn a great deal about filmmaking.

Do casting directors list well-paying and union projects on Actors Access? Absolutely! But whether or not well-paying and union projects are listed, Actors Access is a great resource for the beginning actor, and it is free.

Breakdown Services was created in 1971 to meet the needs of a new age of the casting process. In the early '70s, studios employed a casting department that cast all productions associated with that studio. The casting directors in the casting department were hired fifty-two weeks a year, whether they had a casting job or not. Gradually, studios disbanded their casting departments and finally, in 2003, the last studio with a casting department, Warner Brothers Television, let the casting directors go off to join their "independent" brethren.

Today we are in the middle of another major change in the casting process, as the Internet has come of age. Agents now submit actors via the Internet and casting directors view pictures and résumés online. Breakdown Services allows actors to submit pictures and résumés for its online service called Breakdown Express at no charge. These online headshots and résumés can then be submitted by their agents for available roles. Breakdown Services will provide additional services in the future that will give actors additional opportunities to become more involved in the casting process. The Internet gives casting directors a way to open up the casting process and meet their deadlines and obligations. Hopefully, then, in the future actors won't ask why they can't get the breakdowns, but will instead ask how many pictures they can have scanned in and how long the video stream can be. The Internet has changed the rules and actors have the opportunity to gain from the results of these changes.

Hey, Dad!

Look, I'm Naked!

*N*udity is a subject that comes up every time I give a seminar or do a Q&A at a workshop. There are pros and cons to both sides of the question, which makes for a fine line in giving advice to any single person, but I can say one thing for certain: Never give it away.

Nudity is everywhere you look. It's rampant all over the tube, particularly in cable shows and made-for-cable movies, but even the networks are forced to compete with increasingly sexy shows that aren't afraid of a little skin. So how should an actor handle this dilemma?

The breakdowns come in looking like this: "SAG low-budget, scale plus 10%. Nudity required, looking for a star name." I'm not kidding. This is what agents get all the time (for more about breakdowns, see Chapter 16).

DON'T!

Do not even consider taking your clothes off if you feel uncomfortable. It will affect your performance and that's not what you want. If it is a religious question, do not do it. If you are genuinely worried about your parents seeing it, do not do it. You need to answer these questions for yourself before you are faced with the question from someone you're working with. For every person that says no to a nude scene, there will be dozens ready to say yes, so you should know that if they want nudity in the role, they will have no problem getting it. If you decide you want the role and nudity is required, you should have no problem doing it (that is to say, be clear with your choice to do nudity before you pursue a role that requires it). Don't do it the other way around—that will cause problems for you and for the people with whom you work. In today's market, and with social mores trending as they are, we're going to see more and more nudity.

DO!

Discuss the subject with your agent or manager. If they do not bring it up, you should. They need to know if they should submit you or not and to make sure no one is embarrassed. If you are presented with the prospect of auditioning for a role that has nudity in it and you are on the fence, it is okay to meet with the director or casting director and discuss the role and the way it is going to be shot. Be sure, prior to this discussion, that you have spoken to your agent and manager and you are clear with them. Are they looking for a nude model who can get out a few lines, or are they looking for an actress who will shed her clothes?

I would rather a new, young actress do a nude acting scene in bed with a major movie star than as a guest on several episodes of *VIP*. The movie with the big star will at least be shown at the Writers Guild, the Directors Guild, and the Producers Guild. "Hey, what a body!" might be their first thought, but "Hey, she's holding her own with so and so, let's bring her in for that co-starring role on our next film," might just be their second. That is, of course, if you are a terrific actress and have taken advantage of the right situation.

DON'T!

Do not, under any circumstances, get undressed in a casting director's office. After you have auditioned for the director, if you are being highly considered, you might have to disrobe at some point. Bring someone with you, a friend, or even your agent or manager. Remember, this is not you, it is the character you are playing. All those acting lessons should begin paying off right about now. If you are the sexy type, with a drop-dead body, have some pictures showing your assets. That should work if the director is legit. Don't fall for the old line, "I need to be sure you won't be nervous, so let's rehearse the love scene." Tell the director to bring in Brad Pitt (or your co-star) and you will show him what acting is all about. If standing up for yourself like that costs you the job, well, that's OK . . . or is it? Just be sure to remember that once you give it away it is gone, so make it worth your while.

I am by no means putting down an actor who does "slasher" films with a few nude scenes. There are actually some well-made films in this genre. Many actors and actresses got their start in low-budget, B, T&A, or slasher movies. Some went on to bigger films and greater success. If you hit the big time, no one will care about your early "slumming." If you don't, you can always look back and blame the poor choices you made as a young actor. It all comes out in the wash.

The bottom line: Many big-name stars will do nude scenes from time to time. If *they* will, a new actor will certainly be expected to as well. In this day and age, you should also consider that any performance nudity you undertake will probably wind up permanently archived on the Internet and available to every twelve-year-old armed with a modem and a search engine. It's illegal, sure, but it almost always happens. No nudity escapes the notice of someone, somewhere on the Internet. And this, too, shall only get worse. Get advice from your team, your parents, your religious values, but ultimately it's a personal decision that only you can make.

The Casting Process

From Breakdown to Soundstage

*I*t all starts when your agent gets a breakdown of a television show, feature film, commercial, pilot, or play. This can come from Breakdown Services, LA Casting, or a casting director who might email your agent individually.

The breakdown consists of the following information:
* Production company
* Budget:

   Regular Budget (over $2 million)

   Low Budget (under $2 million)

   Modified Low Budget (under $500K)

   Non-Union (not a SAG-signatory production—if you are a SAG member, you cannot work on a non-union film)

   The budget determines the Screen Actors Guild minimum rate that actors will be paid.

- Director
- Producer
- Writer
- Casting Director

And, of course, the breakdown also contains a list of the character descriptions and names, and whether or not a "Star Name" or a "Name" is required. It will sometimes include an indication of how big or small the roles are and/or how many lines the character has. Finally, there will be a short synopsis of the story.

Next, the agent submits actors for each role by sending hard copies of the actor's photo/résumé to the casting director via messenger or by sending a digital copy over the Internet. Generally, the agent will follow up with a phone call to the casting director to pitch his or her client. This all depends on the relationship the agent has with the particular casting director. Less often, an agent may call a director or producer, if he or the client has a special relationship with that person.

Hopefully, the next step will be receiving a call from your agent for an audition. The first question you should ask is what picture was submitted. Was it the one with the glasses and the bow tie (the nerd shot) or the one where you look like a police officer? This is very important, as a casting director will have agreed to meet you for a certain role and will want you to look like the picture your agent submitted. Your agent will give you a time and place. Do not ask your agent to change the time—this is one of my pet peeves (see Chapter 17)—because many times the casting director has matched up actors. Get there at the

[File 0710f01-iam]L
A SCREAM TEAM PRODUCTION   Producers: Robert Stevens, Jr.
"SKINWALKER: CURSE OF THE SHAMAN"   Writer: Michael Marcelin
FEATURE FILM   Director: Robert Stevens, Jr.
FEATURE FILM   Casting Director: Scream Team Films
NON-UNION   Start Date: August 4-14
LOW BUDGET FESTIVAL FILM   Location: Around Los Angeles
                                            Rate: Non-paying; copy, credit and mileage provided

WRITTEN SUBMISSIONS ONLY TO: SCREAM TEAM
                                            14011 Ventura Blvd., #201
                                            Sherman Oaks, CA. 91423

PLEASE SUBMIT ESTABLISHED AS WELL AS NEW ACTORS.

[BROOKE] Pretty college student [20-25], studying mythology. She is beautiful in the I-don't-know-it-sort-of-way, very determined but likable. Caucasian,,,LEAD

[A.J.] Funky, hip African American college student, 20-25. Cameraman for Brooke's documentary, and comedy relief...LEAD

[SHAMAN] Mysterious middle-aged Native American man (possibly Hispanic) who leads Brooke through events of the curse. Strong man...LEAD

[CHAD] Good looking, egocentric bad boy / jook. 20-25. Caucasian...LEAD

[MEREDITH] Extremely sexy girl, 20-25. Knows how to flaunt it. Comfortable with sexuality. Caucasian...LEAD

[WYATT] Edgy, likable cowboy type, 20-25, good looking in the apple pie sort of way. All-American. Conscience of the group. Brother of Kacy. Caucasian...LEAD

[KACY] Pretty, tough, yet wholesome cowgirl type, 20-25. Sister to Wyatt. Horseback riding experience a plus. Caucasian...LEAD

[EVE] Sexy alternative chick with an edge. 20-25. Tattoos, piercings a plus. Caucasian...LEAD

[SETH] The joker of the group, 20-25. Caucasian. Could be funny looking or grunge. Be creative...CO-STAR

[YOUNG GIRL] 1963. Angry jock. Good looking. Caucasian...CO-STAR

[YOUNG GWYN] 1963. Cute, shy, quiet, type. Caucasian,,,CO-STAR

[YOUNG KENNY] 1963. Marlboro Man, likable guy. Caucasian...CO-STAR

[YOUNG DIANA] 1963. Beautiful bitchy prom queen. Caucasian...CO-STAR

[YOUNG TAD] 1963. Jock trying hard to be cool. Caucasian...CO-STAR

[YOUNG MARIE] 1963. Cute cheerleader type...

A sample breakdown.

scheduled time, unless you are having a lung removed—
even then, get there. The competition is rough and the line
to take your part is long, long, long, so get there on time
(you only get one chance to show you are not professional—no excuses will matter).

The sides (the part of the script that has your dia-
logue) are usually available on Web-based services to
which you can subscribe. If you don't have access to a
computer, then your agent can get those pages for you;
however, this is an extra burden on your agent, so get a
computer and get on the Internet. It will pay off in more
ways than you can imagine (from imdb.com to
craigslistla.com). You can also get to your appointment
early to get the sides, but I do not recommend this. You
need more time than that to prepare and get comfortable
with the lines. You don't necessarily want them memo-
rized, because that might lock you into a performance. You
do, however, need to be very familiar with them. Do your
homework; again, you won't be given more than one
chance to show you are not professional (I can't stress this
enough).

I would suggest giving yourself an extra hour as you
never know whether the L.A. freeway system will cooperate.
The difference between twenty minutes and two hours is
just a little bit of traffic. Be prepared for this; the studios
are scattered from remote parts of the Valley to the
beaches.

When you get there, sign in and hang out. Check out
your competition; you might pick up a last-minute idea.
When it is your turn, get in and get out. Do your lines your
way, then ask the casting director if they would like to see

it another way. Be cordial, and do not under any circumstances have an attitude. If you get called back to read for the director or producer, do it the same way—do not change anything. You are being brought back because the casting director thought you had something, whatever it was, and they now want to show the director or producer what they saw earlier.

Finally, if they like you, and you are their choice, your agent will negotiate a deal for you. That will include salary, billing, and any expenses if you are going to shoot on location (outside Los Angeles). Once that is done, you will get a call from a wardrobe person. They will either get your measurements over the phone, or you will be called in to the studio for a fitting.

After wardrobe, you will get a call from the assistant casting director with your work call, which will include the time and place you are to report for your acting job. When you get to the studio or location, check in with the assistant director (AD). He will send you to makeup and wardrobe or he will show you to the "honey wagon." That's your small (very small) dressing room. Do not go wandering off; stay put. If you have to go somewhere, let the AD know where you are going. You are now a professional actor. Congratulations!

Steve's Peeves

And Other Minor Errata

There is so much to share and so many great lessons I've picked up along the way that it's difficult to fit them all into one comprehensive look at the industry. But I do want to share a couple of "footnotes," if you will. These are things that have been making me crazy for some time. They are mostly little things, but they include one final concern I find quite serious and important.

Let's start with the little things:
- After being given audition information, do not call back several times because you've misplaced it.
- Do not call to say you'll be late for an audition because you are on Wilshire Boulevard in Santa Monica instead of Wilshire Boulevard in downtown L.A. Check your *Thomas Guide* and be sure you know

exactly where you are going when you talk to your agent the first time.

- This one drives me nuts and, believe it or not, it happens several times a week. I begin, "Hi, you have an audition tomorrow at MGM, the casting director is xxxx xxxx, it's for the new series xxxx, your call time is x:xx, the character is xxxx, and the sides are posted online." The client responds, "Can you repeat that? I'm driving and I need to pull over," or "Could you leave that on my voicemail? I'm at the gym," or "I'll call you back later and get the info, I'm on my way to pick up pictures," etc., etc., etc. Why would you let your agent, for whom time is money, drone on when you can't take down the information? Let him know right away if you are not in a position to write something down.

- "Can you change the time of my audition, I've got a lunch with so-and-so," or "I don't have a ride until later, my car is in the shop." No, no, no, no, no excuses, I will not change the time, it's a producer's session and you need to get there when they've asked for you any way you can. This could be your big break, and you could blow it by being late or by changing your appointment when they need to match you up with other actors. Take a bus, hitchhike, I don't care . . . just get there. This is a business, and it only takes that one break. Treat every opportunity as if it were *the* opportunity, or you'll be one of the millions who were "that close" and never even knew it.

- Your agent tells you he's out of pictures and it takes you two weeks to find the time to get new ones to him. Please, people, if you can't see how that kind of behavior spells doom for the aspiring actor, don't even think about wasting my time.

• • •

These are just a few of my minor peeves, and believe me I could go on and on. This is what drives your agent to adopt a specific regimen of Prozac and vodka. Right now, just thinking about it is making me wonder why I still want to be an agent. But the one thing that concerns me more than all the little screw-ups is the actor who is bitter at the opportunities he (she) feels have been denied him (her). If you just want to be a star or just want to be famous, don't waste your time acting—or waste anybody else's time with your acting for that matter. There are so many better ways to achieve notoriety with much less competition. Only act if you love it, and if you do love it, act like you love it.

There is a great parallel between the acting and rodeo professions that I think might illuminate the situation new arrivals in Hollywood encounter. I'll start by putting it this way: If you don't enjoy riding large, furious mammals that would like nothing more than to grind you into the dirt, don't go into rodeo. The same goes for acting. If you don't love acting, don't become an actor. That's not as simple as it sounds. If you don't want to act for the love of it, if you won't work in small under-99-seat theaters for a pittance, you probably shouldn't take another step in that direction. That is not to suggest that you should accept that situation without striving for more. But if you don't love it enough to take it any way it comes, it will eventually disappoint you. But here's another twist—if you don't love acting you can still have a very rewarding career in show business.

The rodeo grounds are littered with failed rodeo stars, some of whom loved it and some who were lured by other aspects. It's the same for acting. However, if you love acting and pursue it from your heart, it will reward you for the rest of your life—whether you "make it" or not. The same is true of rodeo. No one is going to show up at the rodeo grounds, hop on the Widowmaker, and eight seconds later be a rodeo star. You're not going to wake up one morning, walk onto the lot at Warner, and find you are the next Anthony Hopkins, Tom Cruise, or Meryl Streep. You must love it, you must have a burning passion to ride the bull even if no one is there to watch.

An actor typically trains for years to learn the craft, and beyond all that training must have an additional something they usually call "it." A cowboy will train physically and mentally for years. Couple that with some natural talent and a little good luck and maybe he'll be fortunate enough to scratch the top twenty. So few will make it. If you don't love the getting there, you are lost.

Both actors and rodeo stars need to join an organization or union, and both need something to fall back on when the times are tough. Both need to deal with the public and, perhaps surprisingly, learn the business end of their chosen profession. There is a very small percentage of either that will make a living at their chosen profession, but A) you must always pursue your dreams, and B) the ways of the universe are mysterious. You'd be surprised how many paths there are to the hallowed ground of a fulfilling and rewarding career, particularly in show business.

The only real difference is that a rodeo cowboy is probably looking at five years tops, and though the average career of an actor might be ten years, the ones that love it will be rewarded by acting in some capacity for a lifetime. While the rodeo star faces injury and sessions with the orthopedic surgeon, the actor faces stress, depression, and sessions with the psychiatrist. It boils down to this: Do you have talent, and do you have the guts for all the rejection and temporary setbacks? Do you have the heart to follow your dream no matter what?

DON'T!

Don't imagine what it will be like when you're a star, move to L.A., call yourself an actor, and then wait and hope for something to happen.

DO!

Make your life work. If you are spending today as a sacrifice to an imagined future happiness, you are setting yourself up for a big disappointment. This is the approach most actors take. What you need to do is create a life that makes you happy and pays the bills on a day-to-day basis, then add acting to that recipe as much as you can. If you cannot live without acting, then the life you create will include acting, studying acting, and enjoying acting. If you do nothing more than a few plays in Equity under-99 theaters with your friends, you will still be happy, because your life will be good and sustainable, and will bring you joy. If you stop acting or become quite famous it will be the natural expression of you pursuing joyousness. If you are not going to be happy unless you become a star, you are most likely not going to be happy. Vegas will give great odds on your impending misery. It's crucial to distinguish between acting (the work) and being a star. And it's important to remember that discovering that you desire to be in show business in a capacity other than actor is a beginning, not an end.

It's a rough road to the fulfillment of any dream, but don't let anything in this book discourage you. That is definitely not my objective. I hope I have given you some understanding, so you can go forward with open eyes and mind and an expanded idea of the opportunities that lie before you. I wish you the best of luck, but remember: you can make your own luck too.

Postscript

P.S.—Who the Hell Am I?

*I*n 1973 I heard that George C. Scott was going to star in a movie directed by the great Stanley Kramer. The movie was to be called *Oklahoma Crude*. Since I had a connection with Mr. Scott, having cast his first directorial effort, *Rage*, I called Mr. Scott's agent, Jane Dacey, and asked her if she could put in a word for me regarding the casting of this new project. A call was made, and I got the opportunity to meet with Mr. Kramer, who hired me to cast the film.

Several weeks later, Mr. Kramer took me to lunch at the famous Brown Derby on Vine Street, just south of Hollywood Boulevard. About fifteen minutes into our lunch, George C. Scott walked into the restaurant. You could hear a pin drop, as it was rare to see this Academy Award–winning actor in Hollywood. At that time he was one of the most respected talents in show business, having recently won an Oscar for his portrayal of the general himself in the movie *Patton*. He did not notice us. Mr. Kramer, who had never met his star, asked me if I would introduce him. I got up and headed to Mr. Scott's table, and I could feel every eye in the room on me. I'm sure they were wondering, who is this young man sitting with Stanley Kramer and walking in the direction of George C. Scott?

Mr. Scott seemed pleased to see me, and gestured for me to sit down. After some small talk about how our wives were doing, I told him that his director, Stanley Kramer, wanted to meet him. He just stared at me, and I knew what this look meant: He wanted his director to come to his table. It was an old Hollywood power play. Having built a good relationship with Mr. Scott, I felt I could voice my opinion. I told him I thought he should go to Mr. Kramer's table and say hello. Hell, Mr. Kramer was paying him a whole lot of money, and besides it was Mr. Kramer's town. Mr. Scott was from the Big Apple, just visiting. "Come on George, put the guy over," I said, and he seemed to like the way I phrased it.

We got up and Mr. Scott followed me; all eyes riveted on us as we got to Mr. Kramer's table. Handshakes were delivered all around, and after some talk about the rest of the cast Mr. Scott excused himself. So there I was, having introduced the great director and producer Stanley Kramer to the great actor George C. Scott, and it struck me that twenty years ago I was a little kid selling newspapers on the corner of Hollywood and Vine, not more than a few yards from where I was now sitting.

This type of "Hollywood" behavior is not the point, but it is definitely part of the landscape. Occasionally we get these moments, where large sections of our life seem to come into focus. What did it mean to me to be a part of this classic Hollywood moment, in light of the dreams and wishes of a kid selling newspapers? There's a part of every aspiring actor either waiting for this kind of recapitulation or appreciating its having just occurred. The key is to not let it be about that. The experience will not be diminished in any way when it does happen. Perseverance is the key. My goals have changed considerably over the years I've spent in this industry. When this moment came for me, I was neither an actor (as I was when I started) nor an agent (which I've been for the last thirty years). I was a casting director. Being a casting director was one of the extraordinary steps I've taken along the way, and one I couldn't possibly have imagined while standing on that corner hawking newspapers.

I had another experience that struck a similar chord. Sometime in the summer of 1968, I took a two-week vacation. I put on my old cowboy hat and boots and headed to northern Nevada. I had just purchased a bright red Mercury Cougar sports car and thought this would be a fine way to break her in. Somewhere just north of Bishop, California, I stopped for some gas. In those days you didn't pump your own gas. As the attendant approached, I was struck by his appearance. He was a tall, lanky man, older and missing his two front teeth. But he had a smile that would warm the heart of a rattlesnake.

He asked me if I would like him to "filler-up." "That would be great," was my reply. As he was cleaning my windshield, he kept looking at me, shaking his head, and pointing his dirty, gnarled finger at me. He kept mumbling something like, "Yes sir, I know who you are. You bet, it's you, it sure is." He came around to the driver's side and said, "That'll be ten dollars."

I gave him a twenty and he just stared at it. He even rubbed it once or twice. "Thank you sir, yes sir, I know who you are." He walked back to the old, and I mean old, building with what looked like an apartment above it. Within minutes he came out with three ragtag kids and his wife, a forty-something plain-looking woman holding a toddler, and what appeared to be a grandma and grandpa. He was pointing, and they were all waving. As they came closer, their smiles got bigger. The old man came around and handed me my change. "Hey, everybody, I told you it was him, you know Poppa don't lie. You all say hello now." They all said hello and kept smiling and waving, with the exception of the toddler who began to cry and fuss.

I thanked him. And as I got ready to speed off he said once again, "I knew it was you as soon as you pulled in here, I know who you are." I couldn't leave without asking him who he thought I was. His answer, "Oh, hell, mister, you know!" With that, he walked back to his family. As I drove off, I could see in my rearview mirror that they were all standing together like a family portrait, just grinning and waving good-bye.

I don't know if this story means anything to anybody, but on that summer day in 1968 I knew who I was even if nobody else did. There have been many other times in my life when I was floundering and I didn't know. I loved that experience with the old man because we both knew who I was, at that moment, even if he and his family were entirely wrong. You too need to know who you are. Just be yourself, and there will be many others along the road who will know you or think they know you. And, in the end, isn't that the experience we are all waiting to embrace? Knowing . . . it's the opposite, or perhaps merely the other end, of dreaming. But both are merely the side effects of being, which is what you need to figure out how to do in this industry. Welcome to Hollywood!

Appendix A: The Letter Hall of Shame

On the following pages are a few examples of the *wrong* kinds of letters to send to an agent.

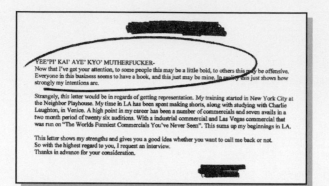

YEE'PI' KAI' AYE' KYO' MUTHERFUCKER-
Now that I've got your attention, to some people this may be a little bold, to others this may be offensive.
Everyone in this business seems to have a hook, and this just may be mine. In reality this just shows how
strongly my intentions are.

Strangely, this letter would be in regards of getting representation. My training started in New York City at
the Neighbor Playhouse. My time in LA has been spent making shorts, along with studying with Charlie
Laughton, in Venice. A high point in my career has been a number of commercials and seven avails in a
two month period of twenty six auditions. With a industrial commercial and Las Vegas commercial that
was run on "The Worlds Funniest Commercials You've Never Seen". This sums up my beginnings in LA.

This letter shows my strengths and gives you a good idea whether you want to call me back or not.
So with the highest regard to you, I request an interview.
Thanks in advance for your consideration.

Do not use profanity, even in jest.

The Stevens Group
14011 Ventura Blvd., Ste.201
Sherman Oaks, CA 91423

I am looking for a partner. I am a diligent, persistent, imaginative actor and I want to
work with an agent who has that passion as well and can represent that. I love to
perform and I have done quite well on my own in my short time in Los Angeles. I
have starred in 6 plays and two short films, as well as small parts in television series
and a feature film. I am also a board member in Son of Semele Ensemble, a critically
acclaimed theater company in LA, and recently finished renovating our new theater
space in Silverlake. I have put in long hours of study and work and I am ready to
grow and succeed professionally. Working with you, Mr. Stevens and Mr. Stevens
Jr., I believe I can achieve that and more. Thank you so much for your time and I
look forward to hearing from you,

Sincerely,

I know that I don't have a lot of credits – yet, but to act is what
I'm supposed to do! Therefor I need an agent who's excited about
what I bring to the table, someone who's got "pull" in the industry
and is ready to go with me on a journey towards a successful career.
I believe that someone is YOU!
Thank you for your attention, and I wish you a beautiful day.
My phone number

Sincerely,

*Asking too much from someone who has no idea what
your talent is or who you are is a real turnoff!*

The Stevens Group
14011 Ventura Blvd.
Suite 201
Sherman Oaks, CA, 91423

Attn.: Steve Stevens Jr.

I'm an actor new to Los Angeles, and I'm up to making a difference in the world through my work. They say everyone has a story to tell. I believe I have something people want to hear. I thought I would introduce myself to you, in the hope that we may work together.

The general quality I bring to my characters can best be described as, 'powerful vulnerability'. In other words, I excel in the complicated roles filled with a mixture of seeming contradictions – sensual innocence, fragile strength and graceful dynamite. Inside of that, there are several character types available to me. I can be sillier than I look, or at home with the gods. I can show people that granite can be fragile too. If I am as successful as I hope to be, the kinds of roles I'd look for would be similar to those played by Jennifer Connelly, Liv Tyler, Diane Lane and Natalie Wood. I share with these accomplished actors a stoic kind of quality that creates touchingly human heroines.

I'd like to set up a time to meet with you to discuss representation. I can be reached directly at ▮▮▮▮▮▮▮▮▮▮ inclined, I know I would be a credit to your office. Thank you for your time and consideration.

Do not compare yourself with great actors or stars.

To:	Theatrical representatives
Re:	▮▮▮▮▮▮▮▮
Purpose:	**Seeking motivated theatrical agent at a quality agency** to help promote and propel the career of a diverse leading man on his rise to success.
Enclosures:	Headshots, Resumes (Please review for your information)
Addit'l Information:	- Great Credits / Quality Shows - Constantly Working & Growing as an Artist - Won Acting Competition for Top New York Casting Director - Appearing in Upcoming Showcase at The Odyssey Theater (in Aug. - details TBA)
Upon Request:	Excellent Demo Reel – (soon to be updated w/ most recent work) Industry Referrals and Recommendations
Note:	Okay, so it's not your typical "seeking agent" cover letter...but it got your attention! ▮▮▮▮▮▮▮▮▮▮▮▮▮▮ love to meet you in person and see if we can work together to find success as a team.

Don't seek and/or ask for special motivation or make outrageous demands right up front.

Appendix B: To Intern or Not to Intern

Learn Production from A to Z: Become an Intern

Being an intern can teach you everything you would want to know about all the different sides of production. It is the lowest job on the totem pole, but the opportunity is tremendous.

Intern on a festival film, a non-union film, or a modified low-budget film, and I promise you will come out with more show business knowledge than you can acquire at any college. You will see close up and in person the good, the bad, and the ugly.

For your first time at an internship, you should offer to take over craft services. You handle all the food, treats, drinks, etc. That way, you become everyone's best friend because you're in charge of all the goodies. Actors will vent to you, the crew will complain to you; in other words you'll be right in the thick of everything. Conversations between the producer and director, and other important people, will take place right in front of you . . . because basically you do not exist. You could write your own book on the *DOs* and *DON'Ts* of a production after an internship like this.

If acting is still your passion, then intern at a talent agency and learn firsthand what it is like to be an agent. Discover what an agent deals with on a day-to-day basis. Then intern at a casting office and see what it is like behind those closed doors. Wow, you spent all those years in college learning your craft and you didn't get one course on how to deal with the reality that this is all just a business like any other. With the exception of the strange aspect of celebrity, it's merely dollars and cents.

Being an intern is hard work and there is no pay. If you get lucky and intern for a person who is willing to teach you, and not just use you as a personal slave, you will be better off for taking the time to do this. When you get your break, you will definitely be grateful that you took a turn as an intern.

Appendix C: Et cetera

L.A. Casting

Non-represented? Don't have an agent yet? LA Casting Network will give you the opportunity to be seen by some of the top commercial casting directors and talent agents in Los Angeles! When you sign on with LA Casting Network you will save both time and money.

$10.00 a month lets LA CASTING NETWORK help you be searched by casting directors!

Now Casting

We're changing the way actors get acting jobs in Los Angeles, and you can be a part of it. For the first time ever, you can instantly submit to high-profile SAG projects created by casting directors who want to see actor submissions. Actors have been called in and booked jobs through our notices.

Now Casting "notices" allow actors to submit electronically to projects created by casting directors on our Web site. They are a part of our Professional, Marketer, or "Works" packages. With a click of a button you can instantly submit to our Now Casting notices. Submitting is fast and easy:

- *Filter projects that don't match your profile with the click of a mouse*
- *Select a project to view roles*
- *Click on the role name*
- *Select your headshot (from up to six photos)*
- *Type in a short note*
- *Click Submit!*

You've just submitted your headshot, résumé, six photos, bio, and demo reel for consideration. It works instantly with no stamps, no envelopes, and no delivery services.

Academy Players Directory

And, as you can see from the following press release, the Academy Players Directory is a part of the special marriage:

January 22, 2002
FOR IMMEDIATE RELEASE

CONTACT: John Pavlik—(310) 247-3090 "Academy Players Directory" Opens in Hollywood Today.

Beverly Hills, CA—The "Academy Players Directory," the casting bible of the industry since its inception in 1937, opens its new offices at 1313 Vine Street in Hollywood, today (Tues., 1/22).

The "Players Directory" is open for business once more in Hollywood for the first time in 56 years. The "Directory" began in

Hollywood 65 years ago, and remained in the community until 1946, when the Academy made its move west to the Marquis Theater at Melrose Avenue and Doheny Drive.

The "Players Directory" will occupy approximately 5,000 square feet of office space on the ground floor of the 118,000 square-foot building that once was the home of the Don Lee-Mutual Broadcasting television studios.

The remainder of the facility will house the Academy Film Archive, which is expected to move sometime in May, and also will provide additional archival storage space for the Margaret Herrick Library.

Renovation work began shortly after the Academy purchased the building in May of this year, and is expected to continue through 2002.

"Our new space on Vine Street will allow the Players Directory to even better serve the acting and casting community," said Keith Gonzales, editor of the Directory. "It's more centrally located, and with on-site parking and ground-level offices, it will be much easier for actors and their agents to come in to conduct any necessary business. Plus, the space allows us to have several work stations for actors who don't have Internet access at home to come in and update their listings."

About the Author

"Extra! Extra! Read all about it!" shouted eleven-year-old Steve Stevens in 1950 as he sold newspapers from the most famous street corner in the world: Hollywood and Vine. At the same time he was establishing himself as an actor appearing in early television shows like *Joe Palooka, The Lone Wolf, Capt. Midnight,* and *Fireside Theater.* By the time he was eighteen Steve was considered a top teen actor with many guest-starring credits behind him.

Through his fifty-some years in show business he has been an actor, a top casting director, a producer, and for over thirty years a Screen Actors Guild franchised agent. He and his partner, Steve Stevens Jr., represent veterans and newcomers at their agency, The Stevens Group, in Los Angeles. After all these years, he still feels that there's no business like show business.